ABOVE THE PAVEMENT—THE FARM!

INVENTORY BOOKS

Series edited by Adam Michaels

01 STREET VALUE: SHOPPING, PLANNING,
   AND POLITICS AT FULTON MALL

02 ABOVE THE PAVEMENT—THE FARM!
   ARCHITECTURE & AGRICULTURE AT P.F.1

www.inventorybooks.info

ABOVE THE PAVEMENT— THE FARM!
ARCHITECTURE & AGRICULTURE AT P.F.1

EDITED BY AMALE ANDRAOS & DAN WOOD

PRINCETON ARCHITECTURAL PRESS
NEW YORK

6	**FOREWORD** **ABOVE THE PEOPLE—** **THE MEADOW, GARDEN, AND COW!** **Fritz Haeg**
12	**P.S.1 BECOMES P.F.1** **Interviews by Linda & Ed Wood,** **edited by Heather Peterson & Dan Wood**
166	**THE ARCHITECT'S FARM** **Meredith TenHoor**
190	**AFTERWORD** **PRAGMATOPIA?** **Winy Maas**
196	**APPENDIX A** **WARM VEGETABLE SALAD** **Michael Anthony**
202	**APPENDIX B** **URBAN CHICKENRY** **HOW TO KEEP A CHICKEN COOP**
204	**P.F.1 CREDITS**
207	**IMAGE CREDITS**

FOREWORD

ABOVE THE PEOPLE— THE MEADOW, GARDEN, AND COW!

FRITZ HAEG

One of my earliest and most vivid memories from a childhood storybook is of an old lady entering her house through a door set into a mound in a meadow. Her house was not on top of the earth, but in it. Up above it there was a cow grazing, and the surrounding landscape of forest and meadow gave way to some vegetable plants, an apple tree, a berry patch, and perhaps some chickens. As you turned the page and saw the house from the other side, there were windows peeking out, framed by tomato vines above and tall grasses below. The surrounding wilderness gradually rose up into a mound, merging so completely with the home where the old lady lived that it would be impossible to clearly define the actual borders of the house. On another page you could see her up there with the cow—milking, gardening, tending—on what one might call, depending on the point of view, the roof. This entire situation—for it was a situation, much more than just an image or even a design—thoroughly captured my imagination. Even at that young age, maybe five or six, I knew this was totally crazy in the most promising way. This was different from any other living situation I had seen before. There was the mystery of it, the modesty, the subservience to the land, its casual nature, and the complete welcoming in of the wild world around it—plus it was ridiculous and fun. The possibility of what this represented had a profound effect on me, and the revelation of this house within the meadow must have also had some effect on my future interests and work.

    The English root of the word garden is *gart,* meaning enclosure. The enclosure serves to keep the cultivated human space inside protected from the wild space outside. It is a space of protection but also of balance between human need and natural resource. Unlike the farm or the designed landscape, it is inherently modest in size so that one person or family unit may tend it. The garden in its richest sense does not distinguish between the productive and the ornamental; rather, these are two ends of a continuum and in some cases are indecipherable from each other. Such a garden is both a negotiation—to determine what sort of sustenance can be achieved from the land in a particular place—and a human space that we occupy, live in, and derive pleasure from. Now that we have dominated and transformed the entire planet, and the wildness that we once protected the garden from no longer exists in its pure form—contemporary gardening battles with wild hedgehogs, rabbits in suburban subdivisions, and urban rats notwithstanding—how should our idea of the garden evolve? If a garden is an enclosure, where do we draw these lines today? What is inside and what is out? Or what if that line is gone and the garden emerges from everything around it—undefined, unprotected, a landscape that goes wherever we are? The lost wilderness and the protected garden would be replaced with a fusion of a cultivated wilderness and a merging of those two very aspects that the garden enclosure previously sought to isolate. The precious but fortressed, controlled garden and the untouched wild would be replaced by a continuous landscape of pleasure gardens—surrounding and above us—that also happen to produce food.

  Growing up in suburban Minneapolis in the 1970s and '80s, I became aware of local architectural responses to the energy crisis and in particular the developing interest in earth-sheltered buildings. At my local library

I discovered the book *Earth Sheltered Housing Design: Guidelines, Examples, and References* written by the Underground Space Center at the University of Minnesota and the Minnesota Energy Agency, and it became my favorite book at around the age of thirteen or fourteen. This rather dry document of charts, section drawings, and data never seemed to acknowledge the fantastic side of the story they were telling, and the authors did not seem to get very excited about what could happen with all of that land on top, but they did validate the idea of living under the landscape as a viable possibility. In my northern climate with harsh winters it was all about going underground to moderate the extreme temperatures and reduce energy costs, something that was a dim and distant goal to a young teenager.

  A few years later I discovered the books of Malcolm Wells, the architect who was baptized in the modern idiom and continued down a corporate path for a while, until inspiration struck, and like a religious zealot he spent the rest of his career advocating underground and earth-sheltered architecture. He designed a few examples, including his own underground solar residence in Massachusetts, which was a significant laboratory for his ideas. Unable to convince many clients of his vision of spaces covered with earth for people to inhabit, he spent most of his energy proselytizing through his books, over twenty-five titles in all since the 1960s. He celebrated the romantic and pragmatic alike with evocative, technical prose, illustrated with lovely line drawings of his architectural creations peeking out beneath layers of plant life. The vision first offered to me in a children's book had come to life here as a possible future for all of us. He proposed a new world order, where we would go down and the plants would go up. Later, Wells would expand his dream to include not just homes, but airports, shopping malls, and offices—everything and all of us would go underground. There is poetry to his completeness of vision, which captures the imagination, but there is also something uncompromising about it, like the emergence of a lost dictatorial modernist voice, reappearing in a new way and triggering a grand reconsideration of the nature of urbanity. How does our idea of a city change if we cannot see it from above? What happens to the identity of a people if their landmarks are not buildings, but trees and gardens?

In Robert A. Caro's 1974 biography of the greatest builder of parks and recreational spaces of the twentieth century, Robert Moses, we begin to better understand why the green spaces in New York, and in cities across America, look the way they do today. On the surface, here was the ultimate friend of greening the city. For decades he devoured land, both unclaimed and occupied, in the boroughs and on Long Island, and transformed one grand parcel after another into an impossibly vast system of giant recreational spaces, parks, pools, and beaches, the scale of which was unprecedented in any U.S. city. Where large tracts were no longer available in the densest, least-green parts of the city (and also the least prosperous areas, which he disdained and virtually ignored), he even started to acquire smaller vacant lots to transform into neighborhood pocket parks. This never amounted to much, as they were deemed an inefficient use of his time in comparison to the scale and breadth of his other, more visible public works. It took proportionally more time to plan and implement a small park, and inversely less fanfare and public acclaim. He was not interested in fostering a dialog with the local community, as would be required for a small neighborhood park. This famously land-hungry man, who would send his team to sniff out any unused parcel

to impose his vision upon, would turn down offers of free
land that were too small and not worth his trouble.
He believed that three acres was the smallest that could
be controlled and managed as a park. Once you have
to start listening to people, things get complicated and
inefficient. You have to find out how they live and what
they want. It also comes down to an issue of scale
that continues to plague our cities, our planners, and
our architects, as well as ourselves to this today. We still
confuse size with significance. How would our cities
be different today if Moses had channeled just a fraction
of his vast resources and brilliance away from the
megaprojects and toward a network of small green spaces
that anyone could walk to or implement for themselves
in their existing locations? Whose job is it to make sure
that each one of us has immediate visual and physical
access to an open green space? Whose job is it to
make sure that every resident of a city has access to
affordable, chemical-free, fresh and local produce?
I have come to believe that no one will ever be hired or
formally assigned to take on the most vital problems
and meaningful inquiries about placemaking for people
today. The most promising and humane possibilities
will come from the collective, unpaid imaginings of
wandering, curious, and self-driven individuals exploring
the edges of what is acceptable at the time, creating
possible alternative scenarios in a much more powerful
way than any hired professional or singular Robert
Moses could ever do on his own. His legacy is among the
most visible approaches to placemaking in the twentieth
century; in comparison, perhaps, it made the old lady
living under the meadow all the more shocking to me.
It was such a contrast with everything else that I knew
about home and city, in my suburban environment at the
time, taking the familiar and shifting it in radical ways

just by putting dirt and life on top. I did not understand that proposition of her life—under the plants and animals in the meadow—as an architectural project, or a literary conceit, or a hippie lifestyle, or a landscape strategy, or a futurist fantasy, or a primitive step backward. I simply saw it as an unexplored, possible way of living—a parallel reality not chosen.

    The old lady living under the meadow, her vegetable garden, and her cow were the first things that came to mind when I saw WORKac's Public Farm 1 installation. Here at last were the visionary architects turning their skills toward a complex living (and alive) situation, not just a design or an image. The inverted V shape of the gardens floating up above was like the inverse of the mound in the meadow, but the grand gesture of the garden coming down to greet us from above felt the same. P.F.1 is not a solution to a problem nor a literal vision of a possible future. That would miss the point and underestimate the project's significance. It is a handmade piece of pragmatic poetry. It's a complex living situation, responding to the larger living situation of the city and the global networks that it exists within. It is alive, both in the lives of the plants that it supports and in the lives of the many sorts of people that converged on a conversation to dream it up, build it, and tend it. The typical team that an architect is used to shepherding has at P.F.1 expanded to include an exciting breadth of human endeavors: the farmers and the engineers, the art curators and solar experts, the college students and experimental soil companies, the graphic designers and chicken handlers. Ultimately I experienced the project as a great big invitation to leave buildings behind, to participate in our landscape, and to climb up.

P.S.1 BECOMES P.F.1

INTERVIEWS BY LINDA & ED WOOD
EDITED BY HEATHER PETERSON & DAN WOOD

---

the competition

Dan Wood: It seems like a dream now, but for a time in 2008 we had more than 150 people involved in a project called Public Farm 1. It was a working urban farm made out of cardboard tubes installed at the P.S.1 Contemporary Art Center in Queens, New York.

Amale Andraos: We really want to work across scales, and we're especially interested in thinking and working at the scale of the city. WORKac started out small, building mostly interiors in New York. Actually, our very first project was a doghouse—we designed a prototype for the "urban dog" with a treadmill and three video screens to provide rural experiences for those long days cooped up at home. Gradually we got bigger projects, like Diane von Furstenberg's headquarters, and public work in New York, and we began thinking and writing at a larger scale.

Dan Wood: That doghouse project stays with us, though. We love the idea that there can be synergistic relationships between things mostly thought of as opposites: city and countryside, nature and artifice, surrealism and pragmatism.

Amale Andraos: We have been teaching at Princeton University since 2005—studio and seminar classes about the possibility of combining ecology and urbanism, and trying to create a new model for cities based on ecological principals. As part of that research, we began studying visionary urban projects throughout history, looking at people who were thinking about cities in entirely new ways.

Dan Wood: We were in the middle of that research when we received the email saying that we had been invited by Stan Allen, the dean of Princeton's School of Architecture, to put together a portfolio for consideration for the Young Architects Program at P.S.1.

Barry Bergdoll: I am currently the Philip Johnson Chief Curator of Architecture and Design at MoMA. The Young Architects Program was started under the previous chief curator, our dear friend Terry Riley. The director of the P.S.1 Contemporary Art Center, Alanna Heiss, and P.S.1's chief curator, Klaus Biesenbach, had come up with the idea of a summer "beach party" based on something similar they had seen in Berlin. They wanted to bring that energy to P.S.1 and decided to also incorporate an architectural installation. The first year, it was basically a disc jockey stand designed by Philip Johnson. The following year, it became a collaboration between the Department of Architecture and Design at MoMA and P.S.1. Terry Riley restyled it as a competition for young architects. Dan and Amale were the ninth iteration.

Sam Dufaux: It was the second time we tried to do the competition, and there was the question, are we too old to get it—not in terms of age, but experience. WORKac had already built a few larger things. We applied again anyway, because it's such a great project. All of the constraints that you would normally have as an architect are removed, and you are left with only your creative mind.

Barry Bergdoll: To be honest, Dan and Amale were a little further along in their careers than other firms at the time they were chosen. In fact, in the jury we had a bit of a discussion about that, but they did something that was so novel that we quickly forgot all about it.

Sam Dufaux: The competition is a two-round process. You have to be sponsored to be considered, and there are about thirty firms submitting portfolios. Then they select five out of that group to develop an actual proposal and presentation. P.S.1 puts equal weight on the ideas and the technical aspects. You have to show a schedule of construction, a budget—these things. They want to make sure you have it all together. It's a lot of work, but it's really rewarding because in the wintertime you're planning for a summer party.

Andres Lepik: I was hired at MoMA as the curator for contemporary architecture and was expected to bring in ideas about who are the cutting-edge architecture studios. When I first came, I was looking around for interesting, younger firms in New York; Dan and Amale's office was my first studio visit. This was a couple of months before the competition for the Young Architects Program; for me, I liked their work even before they were sponsored.

Anna Kenoff: I discovered WORKac when Amale was on one of my final reviews at Columbia. They are interested in an architecture that's about ideas, and a lot of those ideas become political or about bigger issues that really engage the world. That's what drew me to them. I interned with them for a summer then came back full-time the year before we started P.S.1. Dan and Amale had just come back from Christmas vacation, and we all sat down in January and said, "Okay, P.S.1, what are we going to do?" They said, "Look, we have an idea. The history of P.S.1 has always been this urban beach thing, but we're thinking urban farm." For me, that was like, "Okay, you have my attention." The idea that we could engage these other concepts through the P.S.1 competition and breathe some new life into it, push it a little bit. There were weeks of not knowing what shape the farm would take, but no matter what kind of formal ideas we went through, the goal was always to create a real farm. It couldn't just look like a farm, it had to grow and it had to work. Finding a way to make that happen is really what drove the conceptual phase.

Sam Dufaux: Dan and Amale were talking about doing a farm in a very abstract way. Dan is a self-proclaimed farmer, so he was really excited about it. It was also a good year—forty years after 1968—to reexamine the idea of leisure and the urban beach and to question the relationship between the city and nature. We started the usual process that we do in a competition, which is to test a lot of ideas in drawings and models. We had five people working on it. I wouldn't say it went smoothly, but the ideas got narrowed down quickly to this idea of a farm, and then we looked at many different forms. We were also talking about it with people we work with, like Elodie Blanchard, the French artist who we do a lot of projects with. Everybody knows P.S.1 and has been there in the summer; typically it's a beach theme, so we were kind of testing the farm idea on friends.

Elodie Blanchard: The first meeting I went to, their project was totally different than it ended up. It was like they had five projects in one. In the middle of the space there was a huge piece of fabric to make shade, with some water over here and a little garden over there—a lot of things, but no connection. Basically, I didn't like the project. I said, "It's a lot of fabric, plus I don't see the point. You have this big sail thing, but it just makes a shadow all the time the same. Then, because you have to do a bit of ecology, you put the little garden. Most of the time when I've gone to P.S.1, people just want to get drunk and dance." This was my input. Amale, she's always listening, taking notes.

All the day, I say what I think. I'm not going to do something I don't like.

Amale Andraos: That first meeting with Elodie was a real wake-up call. She was right. It was the wrong direction.

Sam Dufaux: We were trying to think too precisely about how the farm would work, and we were having trouble deciding what it should look like. One day a guy in the office not even working on the project came by with a rectangular piece of foam and just put it in the model of the courtyard. It threw us off because we thought we were further along, but it was so simple and looked good. For a while then it was all about this one rectilinear piece. It was floating above at first, but then we realized you have only one point where you can view the farm, and the dirt would be too heavy for it to span the whole courtyard. Even if it could, you would just have the space of the courtyard below, so the goal became to design something that would create different zones.

Amale Andraos: That single bold gesture was so refreshing! The bridge was not form, just infrastructure, which we liked. We also knew we wanted to work with a grid—the urban grid crossed with the agricultural grid—which would work perfectly with the simple rectangle. But then there was the weight issue and the experience at the ground, which was missing. So first we slanted the bridge to one direction, making it only touch the ground in the small courtyard, but this only gave a singular point of view. The structure also got too high at the other hand. So we broke it in two, like a hat or an arch. Finally, we just turned it upside down. The V-shape was perfect; it did everything we wanted and needed it to do in a single operation. It created different zones, for lounging, for kids, as well as varied experiences and scales. Since half of it was now resting on the ground, it helped with the weight issue too. All that was left was to calibrate everything in relation to the space around it.

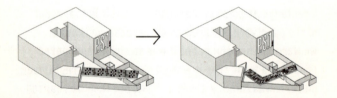

Anna Kenoff: We spent quite a few weeks developing the design. Once we had a direction that we felt good about, momentum soared. As we got more and more excited, we kept adding things to the story.

Mouna Andraos: My sister, Amale, gave me a call and said, "We have this project." The whole three-and-a-half years I was in New York, I kept telling her, "We should do something together. It would be fun. I can help." Of course, a month after I moved back to Montreal, I got this phone call and Amale was like, "So can you come on Monday morning for a meeting?" And I was like, "Not unless I jump on a plane." So we emailed. At first there was the idea of an autonomous energy platform, then the idea of the farm above and these environments, and I really jumped in trying to bring in more human-scaled experiences. We started brainstorming very simple interactions, little punctuations to the whole thing.

Sarah Carlisle: I was a graduate student at Dalhousie University in Halifax, Nova Scotia, which has work terms as part of the school program. I am from a small town in Maine and I wanted to be in a city, so I applied to places in New York. I called WORKac a couple of times, and one day I finally spoke with Dan, and he said, "Can you start tomorrow?" I bought a ticket that night, flew to New York the next day, and started work the following day. I had never been to New York before, so it was slightly intimidating but incredibly exciting.

Anna Kenoff: One thing that was really interesting was working with LERA, the structural engineers. Dan Sesil at LERA knows Dan Wood from over the years, and he loves a challenge. He and Dan Wood have a great rapport; they challenge each other. I think he's been really willing to come in on competitions and help us out because of that.

Dan Sesil: When Dan was working with Rem Koolhaas at OMA, we worked together on the Prada Soho project, and then we did a few things with WORKac: a project in Panama and a fun project in Times Square made out of hay bales. You know, you can find hugely creative people in very modest projects. I've never been particularly taken with size. I guess I view the value of the project in more than just its bottom line. So when Dan called and said, "Could you come over and talk to us?" I went over to see what they were doing.

Anna Kenoff: We had spent all weekend developing this wedge idea, and we finally had a form that we felt good about. It had a pool in the center, it felt right in the courtyard, and we built a model with a grid of squares. Dan Sesil came and we talked about what would be feasible to build this out of. Plywood came up, pieces we could mill and put together like a puzzle, but we were resisting. Plywood felt so toxic and heavy and expensive.

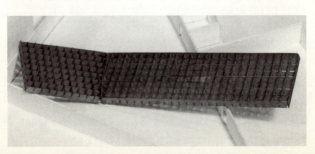

Dan Wood: At the meeting, we were already calculating how many pieces of plywood we would need and how much that would cost.

Anna Kenoff: This went on for about four hours, and just as Dan Sesil was about to walk out the door, Dan Wood said, "What if it's circles? Can't we use those construction tubes?" Some people seemed to know what he was talking about; I know I didn't, but it seemed like the perfect solution. It was a material that we could use in a different way. Paper seemed so much more interesting than plywood. I remember Dan Sesil immediately jumping on board. In hindsight, he probably knew that we were making all of our lives much more complicated by moving to paper, but he likes that kind of challenge.

Sam Dufaux: We started talking to Sonotube, who is a leader in manufacturing these tubes. This woman Dee, in sales, was very helpful. She wasn't a technical person, but she was able to get us samples and basic information on the tubes.

Anna Kenoff: It was immediately clear that the cost was going to be more than we thought.

Sam Dufaux: Dee told us that if we used tubes of three different diameters we could slide them into one another and reduce the cost of shipping by two-thirds. We thought that was brilliant; she's a genius! For them, though, it was like, "How stupid would you be not to do that?"

So we knew we would have to work with tubes of at least three diameters to keep shipping costs down.

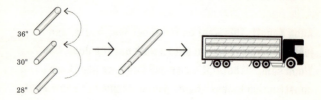

Elodie Blanchard: When you work, things evolve. From the first design, I think they took out a lot of things and kept the garden. That became the main thing. Then they found those tubes. When I saw the next design, the Public Farm, I really liked it because the structure became something else; it had a different purpose. It gave what P.S.1 needed for the summer party, with shadow, water and air, and at the same time it had another function—growing plants. Really great!

conceptual framework and team

Dan Wood: When we started this thing, the closest Amale and I had ever gotten to dirt in New York City was the dust in our apartment.

Amale Andraos: We knew we were going out on a limb by adding this enormous, completely new aspect to the typical P.S.1 installation, and if we wanted to have any chance to win we would have to convince the jury that we had the technical expertise and support to pull it off. We started calling around for advice and to sign people up in support of the project. We called the Queens Botanical Garden and the Horticultural Society of New York first.

Kate Chura: At the time of P.F.1, I was the acting president of the Horticultural Society of New York. We had gotten a call, and I had this conversation with Dan as they were doing the competition. He described the project, and I said to myself, "This is brilliant," because it would bring together urban gardeners and artists. For me, it had great potential to celebrate urban horticulture on a very different level.

Sam Dufaux: There was extensive research into the types of plants we could get, how long it would take to grow them, and what the yield would be. We didn't really want to leave anything open for the jury to question. Basically, we presented it as being the most natural thing to happen in the P.S.1 courtyard.

Anna Kenoff: We always work with models. For the P.S.1 competition we did a general model to show the overall structure, but we also had the idea to do a blow-up of a portion to emphasize the real farm that was happening in these tubes. The two models were one-eighth- and half-inch scale, and we had to search the tri-state area for correctly sized materials to represent the tubes. Some were drinking straws; some were plastic tubes. For the bigger model we found cardboard ones, but they were black, so we had to unpeel the outer wrapper. Everyone in the office had to unpeel sixty tubes at their desks.

There was a lot of testing to make the plant material look good in the model. Haviland was a genius at making plants. He planted the first farm in plastic and clay.

Haviland Argo: I'm from Cynthiana, Kentucky, which is a town of a little more than 15,000 people. I graduated from the University of Kentucky and then got my master's degree at Harvard. Right at the end of school, a classmate who had Amale as a professor told me that they were working on the P.S.1 competition and needed some help for a couple of weeks. It sounded like a great opportunity, so I emailed Dan and Amale, and they said, "Yeah, we need some help." I came at the end of January.

Anna Kenoff: Knowing that Haviland was part of the game put us all a little bit at ease in terms of farming. I remember when he came in the first day and Dan told him that they wanted him to work on P.F.1. He said, "Oh, great, I grew up on a farm in Kentucky." And Dan said, "You're hired. You're staying. You're here for good!"

Melani Pigat: I'm from Canada, from a town in Alberta called Lethbridge. I go to school at Dalhousie University with Sarah Carlisle. Sarah actually got the job at WORKac first. They needed more help doing the models and stuff for the competition, so they asked Sarah if she had any friends that would be interested in working. She called me up and I came. I didn't even talk to Dan or anybody. I flew out on a whim, hoping she knew what she was talking about.

Anna Kenoff: In the last few long nights preparing for the competition, putting together the material, we had to make all of these ideas more concrete. Late one night over pizza we started talking about the shifts in thinking that the project represented—you know, from urban beach to urban farm. Some of those shifts were things that we had observed, like mass industrialization to customization, but the big idea here, a move from suburbanization to rurbalization, was more of a projection, something we would like to see happen.

industrialization	→	post-industrialization
global	→	local
mass production	→	mass customization
suburbanization	→	rurbalization
zoning	→	mixing
segregation	→	integration
free market	→	farmer's market
sand	→	hay

Dan Wood: Because of our urban research, we wanted to both tap into the spirit of some of those visionary projects from the 1960s, these megastructures that were going to provide whole new ways of living, and at the same time to say, "Look, this is 2008, not 1968. We have new ideas about liberation and freedom in the city." One of the rallying cries of May 1968, in Paris, was...*sous*...actually, I can't pronounce it!

Amale Andraos: *Sous les pavés la plage.* Beneath the pavement, the beach. The idea of the beach as a kind of anticorporate, anti-establishment ideal that was just beneath the surface of the city, if they could only tear up all the paving stones.

Dan Wood: Back then it was very "let's hang out and make love," whereas I think our generation is a little more pragmatic. We have these huge environmental issues to confront, and we think of P.F.1 as a kind of first step toward a bigger reorganization of the cities and countryside—what we call "rurbalization," the idea that cities need to co-opt the best of rural life and become more localized and more sustainable. Since we were also doing this by tackling a project with an urban beach history, we came up with...

Amale Andraos: *Sur les pavés la ferme!* Above the pavement, the farm.

Sam Dufaux: Rurbalization, you know, I think the Dutch were the ones who started that whole thing. They have really dense cities and people don't mind that, because within fifteen minutes of the city you have this magnificent countryside. That seems to have influenced the project a little bit. Part of the work of the office is to think about working against this suburbanization of the last fifty years by creating more density in the urban areas and utilizing more countryside for farming—a better balance for everybody.

Barry Bergdoll: Rurbalization is something that has disappeared from American culture. We associate those ideas with victory gardens or hippies or the community gardens that have been under threat of redevelopment in recent years. If you were reading about productive gardens in the city in the last decade, you were probably reading about embattled community gardens. In Europe these things are much more a part of urban culture; all of the unused spaces in the cities are cultivated. That's just starting to permeate into American urban culture.

Anna Kenoff: I don't know if there's one definition for rurbalization. I think, first of all, it's about looking at the city as a set of systems and finding new ways to interweave those systems, rethink them, and make them more sustainable. It's an environmental improvement but also a quality-of-life improvement. We have to propose something new. Is P.F.1 the answer? I don't know, but it's a good experiment.

the presentation

Sarah Carlisle: Dan quoted Noël Coward once—"Work is so much more fun than fun"—and that was a quote that Melani and I included in our work-term presentation. I think Dan and Amale do make it fun. Not only do they make architecture fun, they make work fun. Dan has a very good sense of humor, and their projects are not work, really. It's their life, and it's our lives. I think that by adding this kind of humorous element to their projects, it kind of lightens the intensity of the work.

Elodie Blanchard: They decided they needed a costume for the farmer, and they called me. We had five days. Dan wanted something that you could attach to the tube and put all the fruit and the vegetables in. We started having fun with the thing and made something funny out of it. I thought about a skirt with places to attach all the gardening accessories. It became a really big poufy skirt with metallic grommet attachments and some ribbons, so you could attach it to itself and make a huge pocket; or you could open it—it becomes really long— and attach it to the tubes. They wanted orange; then they thought orange was too much like a prisoner's uniform. Then it was green. It's not really elegant, that skirt, and the fabric is too thin. It's just the idea of it, but it was a good idea.

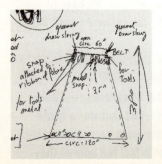

Sam Dufaux: Sonotube was able to give us four tubes in time for the presentation, and we bolted them together in the office and then unbolted them to take them to P.S.1. It was early February and we could not find any good plants, so we went to the Essex Street Market and got some cabbages to make it look like it was all green. We were the last team to present, and we had the room for half an hour to prepare. We brought everything in and put the tubes together like a Formula 1 pit-stop team. When the jury came in, the mock-up was there and Dan and Amale started the presentation. It was this insane performance of Dan wearing a skirt and Amale trying to stay serious next to him.

Barry Bergdoll: The two of them looked like stock actors from the background of a Mozart troupe. We'll just never forget the apparition of this thing. They really staged it. They knew how to do a presentation, to build the drama and not show us the models until the end.

The presentation was exactly like the project—half serious and half tongue in cheek. They understood that it was kind of absurd to propose a productive farm as a setting for a dance party, but they also realized that there was a seriousness about this and a theme that has very much to do with the demands and preoccupations of our time. They hit it right on the mark. We immediately wanted it. We were so surprised; we also wanted New York to be surprised.

Andres Lepik: The jury thought this was the best performance we had seen in a long time. They were so perfectly prepared, and they made it funny. We had so much fun. We all thought we would also have fun with their project in the end.

Barry Bergdoll: The other thing is, over the years, the previous projects become points of reference for the architects. The jury starts to feel, after a while, that they can put the entries into categories. There are the canopies, the tents, the objects, and the benches— but we'd never seen a flying carpet of vegetables before!

Dan Sesil: We knew there would be questions about the safety of an elevated paper structure outdoors, so I came for the presentation, although I think I showed up a little bit late. They asked me one question: will it stand up?

Barry Bergdoll: They were very convincing. They had consulted so many people on how this thing was actually going to function. The only thing that I thought was experimental about it was whether these plants were really going to grow.

winning

Dan Wood: It was eleven o'clock the next day, and everyone knows that when you win a competition they call you right away. Finally, Amale and I went down to get a coffee and be depressed. Barry finally called us at noon, and I was thinking, here we go, the breakup call. It turned out we had won; the *New York Times* was on their way over and they wanted me to put on the skirt! I asked Barry, "What took you so long?" He said he was getting his hair cut.

Barry Bergdoll: I said, "Congratulations, but I have some suggestions." It was my second year in charge of the jury. I felt a little sheepish about giving feedback, but I thought the project was brilliant and that it could get better in its own terms if they were willing to let me play client a little bit. The main problem, I saw, was that they had depicted the farm as viewed from the sky, but everybody was going to see it from the ground, so I suggested that they think about including plants that hung down—moonflower vines and other things. I have a moonflower vine in my garden, and at night it blossoms white flowers that reflect the light of the moon. They would then have a feature related to the night, because the party goes until dark. They were completely open to that. So that was really great.

Sam Dufaux: Right after we won the competition, the article in the *New York Times* brought so many good people on board who were willing to help us, either to work on the project, to donate things, or just share ideas.

Marcel Van Ooyen: I'm the executive director of the Council on the Environment NYC, a nonprofit in the mayor's office. We have four main programs: we build and renovate community gardens around the city, we do environmental education at schools, we educate the community about waste reduction and recycling, and we run Greenmarket, the largest collection of farmers' markets in the country. One of our board members sent me the *New York Times* article. I read it and said, "Wow, this is everything we do. This is gardening, farming, education—all wrapped up into an amazing piece of art." I immediately emailed WORKac, saying, "This sounds like a great project. We'd love to work with you." The rest is history.

Kate Chura: The Horticultural Society of New York runs the GreenHouse program on Rikers Island, where inmates learn to work with plants at our greenhouse and garden there. It's more than vocational training; Rikers is a jail, not a prison, so people are there for short term. The ultimate goal is to redirect a change in behavior so that they can go back and be productive members of society. I thought that if Dan and Amale could grow some of their plants out at Rikers, they would bring this whole other group of people into the project. Our students, who are incarcerated, don't think that there is much potential out there, and with this they would get to be part of an incredible project. It was also an incentive that they could still be involved when they're released. It just seemed like the perfect fit.

Michael Grady Robertson: I work at the Queens County Farm Museum, a 47-acre historical farm with livestock and vegetable production. We're the only working farm of this scale that remains in the city, the last vestige of a long history of farming in Queens. Right after I got this job, a friend of mine emailed me the *New York Times* article about P.F.1. I read it and thought it was a really exciting venture, so I emailed Dan and Amale and said, "Hey, we're a farm out in Queens. We'd be happy to help, you know, propagate your plants and give you any sort of advice or help that we can." They got back to me not too long after that, and we formed a real relationship.

Melani Pigat: So there we were in a huge city, trying to start a farm in the winter. We were going to grow all of the plants in the prison, but the greenhouse wasn't big enough. Michael contacted us; he offered us space in his greenhouse and help to grow the rest of the plants. There's no way we could have done all that on our own. We didn't have the space, let alone the expertise—luckily, he found us.

Michael Grady Robertson: People here were really eager to get involved and be associated with the project. Some of our team have an engineering background and were fascinated in the engineering aspects. Others who have a theater background were interested in the spectacle of it. Our executive director is the president of a museum council, so she was interested in collaborating. The nature of the project made it attractive to so many different organizations and so many people on a lot of different levels. It brought a lot of enthusiastic, like-minded people together.

Marcel Van Ooyen: I don't know that Dan and Amale are an example of every architect you'd work with. I think they didn't have a lot of ego in this project because they had no idea what they were doing, but they were very smart in reaching out to a lot of people and asking for help. They were pooling information from different folks. Any time I would pitch something, they would say, "Oh that sounds interesting, we should check that out." I don't know if that comes from the fact that farming is something they'd never done or if it was just their personalities. Probably a combination of both.

Elodie Blanchard: Dan and Amale are open. You have to be, to do a good project. I don't think you can do a project like this by yourself.

now we have to build it

Amale Andraos: So now we had a big idea and an amazing supporting cast ready to help and a great team of people in the office totally invested in the project. But we had no idea how to actually do it, how it would be structured, how we could organize what we rapidly realized was going to be an enormous amount of work.

Sarah Carlisle: I thought it was interesting to have this urban farm, but it was hard to imagine that it would really happen. It just seemed a bit too "out there."

Melani Pigat: When we did the competition, we just made a presentation book; we hadn't really thought out exactly how we would make this thing stand, how we would get all of these huge tubes of soil up there and all these plants.

Marcel Van Ooyen: My biggest concern was that it was going to fall on somebody's head. As someone who runs an organization and constantly worries about liability, I was saying to myself, what the heck are they thinking? This is going to tip over or one of those buckets is going to come unbolted, but I'm sure MoMA is insured up to the gills, and they're good architects. I guess they know how to build stuff so it doesn't fall apart.

Haviland Argo: There were so many unknowns. It wasn't like we were building a house with 2x4s. We were in completely new territory, building with cardboard. Nobody had drawn up specifications for another project that we could look at and say, "Oh, this is how we do it, this is how it's detailed." We were coming up with everything brand new. That was kind of daunting.

Anna Kenoff: So we called Dan at LERA and said, "Okay, you helped us convince them that this thing would stand up. Now you have to help us figure out how to actually do it!"

Dan Sesil: This was the first time we had worked with a paper structure, but it was more than just about the material. That particular repeater, the tube unit, is a very common construction material used to build concrete forms. To me, that was a part of the interest—

to take something that is otherwise very mundane, very common on a construction site, and assemble it in a way that's more sensitive.

Patrick Hopple: I went to engineering school at Pennsylvania State University and then came to work at LERA. Dan Sesil called me down and said, "I have something I need your help on." I got excited and said "What is it?" "It's a paper project," he said. "A what?" I said, "You're kidding, right? It's indoors at least?" "No, no," Dan said, "It's outdoors." I said "Okay, this is going to be fun." It was one of my first major projects.

Sam Dufaux: When we were planning for the competition, the engineers were very confident with the structure. We showed the sample tubes to them and asked, "Do you think these can work?" They were essentially laughing at us, saying, "These are so strong, there's just no way it's not going to work." Once we got the commission, though, we realized it had to be engineered in a very precise way.

Anna Kenoff: The engineers had a whole perspective about the way things should be constructed—the sequence—and how that should inform the detailing. The original design was a simple grid. There was Row A: one big tube, one small tube, one big, one small; and Row B: two small tubes, then a gap, two small, and a gap. We placed the support columns based on where we wanted them on the ground.

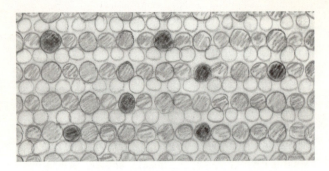

Dan Sesil: The first things we saw were that they didn't have enough columns and the rows weren't a good system; they weren't supported in two directions. Most of what we do are one-off fabrications, every part unique; however, there's a real elegance in creating a premanufactured piece for assembling identical modules. This was a perfect application for that.

Anna Kenoff: At first the engineers were sketching over our A/B/A/B grid, adding a lot of columns and playing with the rows. We went back and forth with them for a week. We didn't want to change anything from what we had presented. Then they came in for a big meeting and said, "Look, we're laying it on the line. You have two big issues to address."

Patrick Hopple: We said, "First, to build this thing, especially with a bunch of amateurs, you're going to have to make a unit that somebody can build on the ground and then hoist up. Second, to make sure it stays up, you're going to have to supply more columns and they can only be so far apart." Also, once we started tracing, we started to see nice patterns in the circles. We started looking at them, saying, "What if we did them like this? What if we did them like that?" We overlaid all the different ideas and eventually started seeing the development of a modular unit.

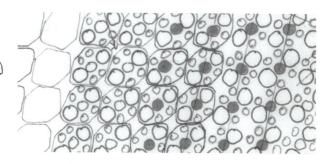

Anna Kenoff: We all sat for hours and sketched and fought until Dan Sesil created the ultimate unit by tracing a version of our first grid, then rotating it ninety degrees, flipping the tracing paper over, and retracing it reversed. That was it!

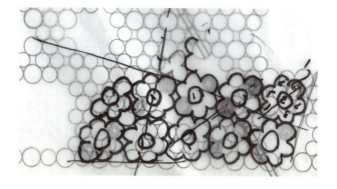

Patrick Hopple: Amale said, "That looks like a daisy. We could have a bunch of these daisies on columns. They will look like flowers!" They took that and ran with it. They set up the grid of columns, and we adjusted and went back and forth.

Anna Kenoff: The grid that emerged had two large tubes and four smaller tubes (two on each side). At the center is the third size. Those were the daisies. Every other daisy has a column at its center. The ones that don't have columns have picking holes to access the vegetables for harvesting. This structural grid also became the planting grid.

Dan Wood: Each daisy held a specific crop: six tubes of beans in one, six tubes of radishes in the next. The idea was that, in the end, you would be able to read all the daisies by the different colors and textures of the plants.

Patrick Hopple: That meeting was tough because WORKac was really set on keeping the design the way it was, but when we talked them through it—explaining that the modular daisy units would basically be the most buildable and affordable—they started to get into it.

Dan Sesil: I've never been one to approach design as a moment of inspiration; I view it as a journey. I also feel that you have to be able to voice your opinions. I can't just be the engineer, where the architect says, this is mine, go off and structure it. To me, it's a conversation, and the projects are better for it.

Anna Kenoff: We researched examples of these tubes being used. Shigeru Ban has used them in architectural projects, but we were working on such a different scale. But then LERA learned that in many of those projects, the tubes were re-supported in some way. The challenge was to make the tubes do the work.

Dan Sesil: The thing that's interesting about the Sonotubes is that they're pre-engineered. You can buy any size, and it's organized to function as a circular tension ring. You pour concrete into it, and the tube prevents it from expanding out. It functions in pure tension. So what we said was, if we could capture that property and use it in reverse—as a compression ring—we could utilize the material better. If you have a circle like this and you push on two sides, it collapses pretty easily. So the daisy was critical structurally to make sure that all forces were uniformly distributed around the ring. The idea of having contact between tubes is very important to the function of the structure. At the same time, the "daisy," from an architectural standpoint, was really playful, given the ideas of the project. There's something fun about that, right?

Sam Dufaux: For weeks we were testing how to get this shape, fit all the tubes in the daisy module, and not have any gaps in between.

Dan Sesil: The compactness of things was really important—that part was quite fun, actually. When they make these paper tubes, they start with an inner ring of a set dimension. The way we varied the sizes—to make the tight, compact fits that we were looking for—was to vary the thickness of the paper wall. If we wanted it a little thicker, we just asked them to lay a little bit more paper on it. While our variable was a coarse fit, we made it a fine fit by varying the amount of paper they rolled around the cores.

Haviland Argo: Even with all of the calculations, we somehow ended up one night realizing that there was still a quarter inch of separation between the actual diameter of one of the tubes and the way we had drawn them to fit together. It doesn't sound like much, but when you multiple it by 352 tubes, it really adds up and could have been a big problem. We had to reconfigure the entire layout. That was something that was, I think, kept from Dan and Amale until we had worked it out. We stayed up all night redrawing everything before telling them.

Patrick Hopple: We then had to take this concept and get it into a 3-D structural model to check all the forces acting on it. This was the first project that we actually modeled from scratch. We made some assumptions and ended up producing this entirely new type of computer program. In fact, our IT department told us the other day that it was the largest folder on the server in the office. I mean, we're talking about other projects here being the Shanghai World Financial Tower, one of the tallest buildings in the world, and its folder wasn't as big as P.S.1's. I was like, "Oops!"

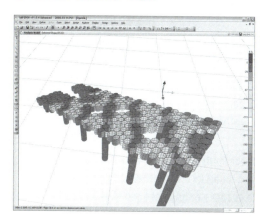

Dan Sesil: We ended up putting plywood disks in the tubes to support the plants, but these also reinforced the structure. Think of the tube as a circle; if you push on two sides, the circle will want to become more egg shaped. To allow the circle to actually carry some type of load, what you want to do is restrain it from bending into that shape. The plywood disks provided that restraint. The picking holes had disks as well, with a hole cut out of the middle. In places where our structural model showed us that the forces would be too great, we added a second disk at the bottom. If you think of the tube in section, when you push on the top, the bottom will want to bow out. What the two disks were doing in those places was forming a gripping couple; while one disk was compressing, the other was pulling. That allowed the tubes to carry more force.

Melani Pigat: Anna had gone to school with some guys who ran a milling company. We got to use their mill basically for free, so that was a huge expense cut.

Amale Andraos: Many of the previous P.S.1 projects were all about using new computer software and 3-D programs to create complex shapes that were then computer-milled out of different materials in an attempt to really harness new technologies in the service of design. The structural engineers did the only real software innovation on our project.

Dan Wood: And the computer milling was essentially reduced to drilling holes and cutting circles out of cheap plywood.

Patrick Hopple: We ran all of our calculations by the engineers at Sonotube and they were still pretty certain that the tubes were going to collapse under their own weight, so we limited the stress and put the discs in to keep it constrained on itself. The discs were a very smart move to counteract those stresses, but we didn't know what it was going to do, really, until we built it.

Dan Wood: Some architects get really into construction—we have friends who build huge parts of their projects in their studios and install them themselves. Other architects do design-build and act as contractors. Amale and I, well, we're not really that handy. That's why we're architects! We're good at thinking about things, drawing things, making nice models…we're just not that good at hammers and saws and building stuff. We needed a dream team.

enter Art Domantay

Anna Kenoff: We were having really intense meetings with LERA, and we were just at the point of panicking about how we were actually going to get this thing in the air. Dan and Amale knew this guy, Art Domantay, from helping with projects for Creative Time, the public art group. Dan went to dinner with somebody one night; that made him think about Art. The next day Dan came into the office and I remember him saying, "Art Domantay is going to save our ass."

Haviland Argo: Dan had worked with him before and was like, "We've got to talk to this guy. He's a real expert." Anna and I were the first ones to meet with him. We explained the project, and I realized instantly that he grasped the concept, the magnitude of it, and how we were going to accomplish it. I felt like something happened. I told Dan, "We have to have this guy. He's awesome. I don't know that we can do it without him."

Art Domantay: I wasn't quite sure about the project. I had just come back from LA after installing another project, and I was pretty tired. I also never really thought I'd work on any of the P.S.1 summer projects. I knew about the previous ones, and it's always been a very daunting process. The compression of deadlines is almost impossible—literally a month or two to build the whole thing—so it was a pretty big deal for me. It took, I'd say, about a week and a half for me to say yes or no.

Anna Kenoff: I remember sending Art a lot of really, really nice emails, trying very hard to get him on board. We suckered him into it. He was...everything. He made it happen.

Art Domantay: I believed in the project conceptually. The first thing that I said to myself was it would be great to be involved in this. That was my gut feeling. I guess I also took a liking to Dan. He was fairly easy going and, at the same time, he and Amale obviously needed help. I've done a lot of big, massive projects with people like the Public Art Fund, with limited funds and limited time. So Dan and Amale's project fit in, pretty much, with other projects that I've done in the past, just on a much bigger scale and with a really compressed schedule.

Sam Dufaux: It's pretty tight—everything in six months. You work one month on the competition, another month and a half organizing your team and project.

Once you start building, you have to make sure it's a really efficient process.

Art Domantay: I talked to a number of professional fabricators who have been my friends for years. I rely on their guidance and sometimes consult with them on projects. All of them said, no, don't do the project. They told me I was screwing myself; I would not make any money, and I would never finish on time. Then I talked to my girlfriend, Marie. She said that I should do it because it's hard, because it's a real challenge. I knew then, deep inside myself, that I needed to do it. I just needed that one voice to tell me that I should.

Amale Andraos: I can't tell you how excited and relieved we were when Art got on board. He was the last piece in the puzzle.

Art Domantay: Dan and Amale thought that they could just build it at P.S.1. I've worked on a lot of outdoor projects, and I knew that in two months' time you're dealing with all kinds of weather that can really slow down the project. I told them that if I did take it on, there were two requirements: one, that we build it somewhere indoors; and two, that I needed to build a temporary ramp on-site to put it all together. That ramp and the space were both going to increase costs, and I knew that every dollar counted. Those two conditions were both necessary safety factors.

Dan Wood: We knew this was only the tip of the iceberg in terms of all the unknown costs and complexities, so we redoubled our fundraising efforts and got some great people on board. The Seed Fund from San Francisco approached us, and Marcel at the Council on the Environment NYC (CENYC) found some of their board members to support the project. But we were scrambling. This went on all the way to the end.

Art Domantay: Just when I started to think about where I could build it, the company on the bottom floor of our studio went out of business. I told our landlords about the project, and one of them became interested and agreed to rent it to us for a couple of months. After officially committing to the project, I started to clean up the warehouse. It was an old sweat shop, 9,000 square feet. Luckily there was an elevator, so after we cleaned it up, everything from my wood shop just went straight down to the bottom floor, including my bathrobe, because I knew I was going to be sleeping down there a lot.

Sarah Carlisle: Melani and I shared a bedroom in Brooklyn with five other roommates. We didn't have heat in our apartment, but we were working a lot, so when we were home, we really just slept. We had each other, and we made it work.

Art Domantay: The crew started small. First I brought in two really great carpenters, Tim Daley and Grady Barker. We figured out what was necessary to get it going. Then we built up the crew. Everyone had done big projects before, but none of these people knew what it was really going to take. Melani and Sarah really helped out. Dan and Amale were busy trying to find money, working on other aspects of the project and running the office. The other architects were dealing with supplies and schematics, getting drawings out. I relied on Melani and Sarah to help visualize the project.

Melani Pigat: Day one, we went out there and I would say we were a little bit shocked, because it was Sarah and me and all these tubes. We put our masks on and started lacquering away.

Sarah Carlisle: At first we thought we had to do one coat of lacquer on the tubes. The very first day, I remember painting the first strokes of lacquer on the tubes and I was so excited. Melani and I were taking pictures: the first stroke! Then we found out that we had to do three coats. Then the sea of tubes was delivered a couple of weeks later. It was just so overwhelming.

Art Domantay: The warehouse was for manufacturing, but it was also for storage. We had three enormous loads of Sonotubes, and they all had to be stored there in the warehouse. We'd take them off the 18-wheeler, and it would be everyone unloading.

Anna Kenoff: The big question was—how are we going to cut these things? We were concerned about this physically, and LERA was concerned about their structural integrity. At one point, we were watching videos of these lumberjack competitions with the guys on either side of a giant tree with double-handled chainsaws. We looked into buying some giant handsaws. We were researching sawmills and huge laser cutters. Then Dan Sesil said, "Let's just come over and see how difficult this is." He came in jeans and goggles, with his own personal Sawzall, and it turned out the tubes just cut like butter. The real problem became figuring out the angles to cut them at to fit them all together.

Art Domantay: WORKac determined that all of the tubes had to be cut at a seventeen-degree angle, but we still had to figure out how to make sure we were cutting them all consistently. That was pretty difficult. I think I spent two days building two giant jigs for cutting the Sonotubes.

Anna Kenoff: Art's warehouse crew, who are really the best at building stuff—there was a violin maker and guys who do the Christmas windows at Saks—couldn't get their heads around what went where. We needed really clear assembly drawings; a team in the office started putting them together. Each tube had four to six 2x4s running vertically, depending on the number of connections to other tubes. These created a gripping surface so the bolts wouldn't rip through the paper. Each 2x4 was Liquid Nail glued and screwed with sixteen screws from the outside in. Each connection had three bolts through holes that had been predrilled in the 2x4s. Each 2x4 had a 4x4 "key block" lag bolted to it to support the planter disks, which were screwed down. The disks were at different heights, depending on their position within the overall plane. Some tubes also had double discs for reinforcement, and the picking holes had discs that were rings. It took forever to figure out the specifications for each daisy, and the pressure was on. Art was frantically trying to get everything going. He needed those drawings.

Dan Wood: Art is an artist as well as an installer, and we always say that this is what made him such a great fit for the project. He was able to approach this messy construction problem with an artist's eye and a huge amount of creativity. The whole warehouse setup was really an ingenious artwork in many ways—part Rube Goldberg, part Tim Hawkinson.

Anna Kenoff: Of course, the whole idea of the prefabricated unit was to simplify building, but the actual geometry had a hidden complexity that required Art and his crew to build intricate jigs for each step of the

enter Art Domantay

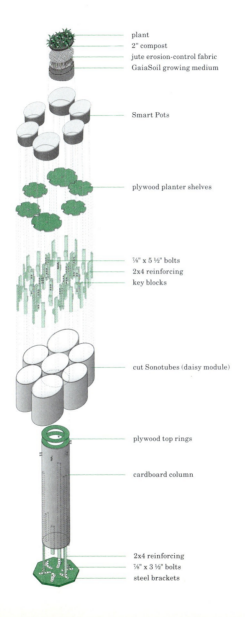

process. Finding a way to keep it simple, I think, was the challenge that Art really wanted to take on. He loves tools and toys.

Sarah Carlisle: First there were the jigs he made to cut the seventeen-degree angle on the tubes. There were three of these, one for each tube size.

Art Domantay: These were not just jigs; each one was like a small locomotive. The guys created a track with wheels all along it. The lightest Sonotube probably weighed about 300 pounds, and they could fit these huge tubes onto this train car, spin it around, and slide it through the jig. That enabled them to cut so fast. It was daily troubleshooting. I told everyone that if you just keep at the pace you're at now, we'll never get it done. After each cut, you've got to figure out how to do it faster and quicker. We had very few extra Sonotubes, so each cut had to be a good one. In the beginning, we could cut maybe two or three tubes a day, but at the end we were cutting ten tubes a day without any mistakes.

Anna Kenoff: Art set up five or six assembly lines. As more and more volunteers showed up, they just kept adding stations. When they came off the cutting jig the tubes would go outside, where teams were constantly lacquering. Other jigs were set up to cut key blocks and 2x4s, drilling the holes where they needed to go. There were just piles of 2x4s all over the warehouse.

Sarah Carlisle: There was a jig for labeling and marking the 2x4s, an intricate system with hinges. Art loves his hinges! They marked the 2x4s at different heights according to where the key blocks needed to be and where the holes had to be drilled for the bolts. Once the holes were marked, the 2x4s were taken to be drilled. There were two drilling crews, making four holes in each 2x4. Each 2x4 was labeled with a number and letter, showing the daisy number, tube letter, and position within the tube.

Anna Kenoff: Once the tubes came in from lacquering, they would gather the sizes they needed and place them on the sloped template to start assembling the daisies.

Sarah Carlisle: There were three main guys working there; they became incredibly skilled. From the outside, it looked really confusing: they were standing on the ramps one minute, inside a tube with this huge wrench the next, then you'd look over and one guy would be on his stomach with his head inside the tube and two drills in his hands.

Anna Kenoff: The columns had 2x4 reinforcements all the way up in four to six places. There was some sort of jig to know where this needed to go, and people were climbing through the tubes gluing and screwing all the way.

Sarah Carlisle: Then there was a jig for the column bases. They had to connect the 2x4s coming down inside the tube to these custom-manufactured L-shaped brackets. Those were wrapped around the circumference of the tube for attaching the columns to the concrete foundations. The jigs made them evenly spaced, depending on how many 2x4s came to the ground.

Art Domantay: What was difficult was that there was no time for testing. We got the drawings from WORKac, set up shop, and started to build the pieces for a mock-up.

But it never turned out. We planned to put four daisies together with a column to see if it was going to stand up, but there was no time. We couldn't break for a single day. In any case, there was no room. We couldn't take it over to P.S.1 because that would take days. In the first three weeks, we only got one daisy built. We needed forty daisies, and we had eight weeks to build them. It was simple math: there was no way we were going to get it done. No way.

value engineering

Sam Dufaux: At some point we realized that we had planned for something that was going to be too big, and we had to scale it down. I was personally really scared that it would lose its substance, because to have these two tilted planes, they had to be big enough to look like planted fields. If we went too small, it would just look like a few rows of tubes. I really was very concerned that by losing the size, we would lose a big part of the idea.

Anna Kenoff: It had gotten to a point where we'd already called everyone we knew for money and donations. Some people were responding; some weren't. We had done enough detailing to really know what we were in for, so we started to rethink certain aspects of the project.

Amale Andraos: It felt like we were cornered. We had a dramatic financial situation, but we were worried that it would look terrible smaller. For us, it was almost better not to build it at all than to make a compromised version. The whole thing was based on the idea of filling the courtyard with a minimal operation. Proportions and size were critical. It looked simple, but it was very precisely designed.

Elodie Blanchard: I would have loved to see it bigger. When I saw the model, I thought it would be so much bigger than the actual thing. I think it was originally, but then it had to be shrunk down because of budget.

Amale Andraos: We reduced it from four daisies wide to three and added the "ground daisy" seating areas. We sent the revised design to Barry and Andres at MoMA. They thought it was okay, but we still weren't convinced. We mocked it up at P.S.1 with stakes and looked at it from the windows above, together with Tony Guerrero from P.S.1. He agreed it was too small, so right there together we decided to add one row of empty tubes along both sides to retain some width and proportion. We quickly moved the stakes and agreed that it looked good.

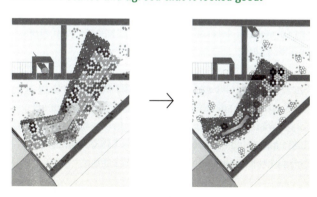

Anna Kenoff: Making it smaller was fine in the end, but it was a scary leap for us at the moment.

Dan Sesil: I was one of the voices at the table counseling Amale that it was going to be plenty big. All the same, it was a relief to go out there once it started to take form and feel good about the size of it.

bolts

Patrick Hopple: We were studying the structure in the computer model as they started building everything in the warehouse. Deflections were color-coded so that we could read the size of the deflections, ranging from blue (which was okay) to red (indicating failure.) When we first ran the program, there was a ton of red. Fortunately with this program, we were able to go to particular pieces of the structure and look at specific information for each node. We could draw a "force cut" through the joint, and the program would say, for example, there's this much shearing between the two joints or this much tearing in the paper. From there we were able to get all of the forces and determine the reinforcements, bolt sizes, and locations.

Sarah Carlisle: We had to buy a really large quantity of bolts, so I was speaking with different bolt companies to see what kind of deals they could give us and calculating costs and shipping. It was a lot—like $16,000 worth of bolts at the beginning! Dan was freaking out.

Art Domantay: I'm just the builder. To some extent I had to worry about the money, but really I think that one of Dan and Amale's most important roles was to sell the project and figure out where the money was coming from. It was such a small budget, you know. It was tiny tiny and everyone knew that. Dan and Amale needed me to construct; I needed them to figure out where my supplies were coming from.

Patrick Hopple: This was not only my first time working as a project manager, but the uncertainty of the material itself also made it extremely stressful. As I worked on it, I got more and more uncomfortable, to a point where I was not sleeping very well. Usually, designing with a familiar material like steel or concrete, you have a sixth sense of how the material will react. You know that you need this many bolts for steel. But here it's a paper structure. I need how many bolts? Really? Is that enough?

Art Domantay: The bolts just kept getting bigger and wider. The engineers wouldn't let up. For a long time there was no compromise at all. That was really difficult for me. I was asking myself, "How are we going to construct this?" We have six weeks left and a couple of daisies done, and we're getting more and more drawings that are more and more complex. The initial glossy brochure that I said yes to turned out to basically be a cartoon of the real project. In the drawings everything was fine, but actual construction was another thing.

With the lag bolts that the engineers wanted, for example, and the direction they wanted the screws and everything to go in, we would have to build the daisy, take it apart, put the screws in, and then build it over again. Every daisy! So it was completely impossible.

Anna Kenoff: There were immediate concerns about the amount of warping and how to make this cardboard structure really stand up while being exposed to so much water and soil. The engineers had said, give us a load and we can handle it. But their way of handling it was a massive number of bolts. When we started running budget numbers, we quickly realized that we were still in serious trouble.

Patrick Hopple: It got to a point where it was very stressful for a lot of us. Tempers flared a little bit. This was such a custom project. Sometimes we got a lot of resistance to our ideas. When you get to something that steps out of the box this far and has so many people thinking on edge, you know, it's pretty tough. But we slogged through. We all worked together, making simplifications. We have a very good relationship with WORKac now. Hopefully it'll never get to that point again, but it was fun along the way, just experiencing something so intense.

Art Domantay: It was an amazing engineering feat, but many times I would say: I just don't know how you can build something like that. I guess it's easy to specify twenty screws on one side of a 2x4, but I would have to say, instead of twenty can we do sixteen? I would have a gut feeling that something would work. I'm not an engineer; I'm not even a big builder. But we had six 2x4s in every tube, seven tubes in each daisy, and more than forty daisies. Four less screws in each 2x4 would save a lot of time and materials.

Dan Sesil: The only problem we had on this job was cost. Poor Dan Wood was trying to pull this off from a financial standpoint and it was not easy. Most of what you saw out there was structure, with the exception of some dirt. A big part of the cost was due to what I was saying they had to do. Some of the push-back from WORKac on this project was simply because they absolutely had to pull it off. And they wanted to know for sure if they could get rid of some bolts or change something. Probably the biggest change we made that helped the structure the most was going to a lightweight engineered soil.

Anna Kenoff: One of the ideas for reducing the structure was to find a way to lower the weight. We were searching and searching. I don't remember finding a lot of options.

Marcel Van Ooyen: The soil was a big challenge. We helped them find an alternative. People in the growing community had talked to us about this GaiaSoil a while ago, but it was designed for rooftop gardens and we don't do that. We talked to WORKac, though, and suggested they check it out. That worked out really well for them.

Soil and soil scientists are amazing. I think we take for granted the intricacies of soil.

Paul Mankiewicz: I am a biologist and was involved in research in the late '70s and early '80s, when there was a lot of talk about farms on rooftops. Soil is sand, silt, and clay, plus organics. Sand and silt are silica dioxide, basically just glass, they don't add anything in terms of nutrient. They form the physical structure that holds the clay and the organics and allow space to hold water and for roots to penetrate. I realized that you don't need something as heavy; you could use anything that had the same general structure. So at one point I was holding a Styrofoam cup and realized that you could probably replace the sand and silt with Styrofoam. Styrofoam was out of favor then because one, it was filling up landfills, and two, people thought it was toxic. But scientists at the time tested it and discovered that the styrene leached from a Styrofoam cup was tiny, insignificant. So, I made a number of prototypes of a soil made from recycled Styrofoam that was headed to landfills. That became GaiaSoil.

Anna Kenoff: Paul was fantastic. He showed up in the office and just started talking about all of his various projects, and I think immediately we just really connected. Paul is this crazy mad scientist and really enthusiastic. He immediately got excited about the project. I think GaiaSoil had been tested on some rooftops in New York, but I don't know for sure if it had grown vegetables before in the amount that we put it to. He came into the office carrying a strawberry plant in his soil to convince us that this would work, that it would actually grow. Some of the other horticulturalists that we were working with were really hesitant about this soil, but it proved out to be a fantastic test for the invention.

Paul Mankiewicz: You know, we've tried lettuce, tomatoes, grape vines—it's been a success for everything. We can grow anything. But when we created it, the faith in urban agriculture, the notion that we should do these kinds of things in cities wasn't widespread. There was a small minority of green people, and they didn't have the capacity to put these ideas into practice. GaiaSoil came out of this attempt to restructure urban ecology and animate sterile buildings with life. It's just that people didn't have the interest and the wherewithal to do it; until the Public Farm, there was no artwork, no architectural enterprise to demonstrate the feasibility of the concept at a wide scale.

Haviland Argo: If we hadn't found the GaiaSoil, I don't think we could have made the project work. The amount of weight we would have had to carry and the cost would have just been enormous. Also, all of the logistics of lifting the soil into place couldn't have worked the way they did. We had looked at using a blower truck and blowing dirt up into the farm, but in the end we were able to do all the planting on the ground and lift everything up. It was much safer and more efficient.

Dan Sesil: It was a child of necessity, right? They were fighting cost. I was telling Dan and Amale that we have to put all of these fasteners in there, saying, "Look, if you put all that dirt up in the air, it isn't going to be light." Well, then they went off and found some dirt that was light!

Paul Mankiewicz: We donated it. We're a small not-for-profit, so it was something of a hardship but we were able to couple it with some other jobs at the same time. It was a lot of product, you know, but I was happy to give it to them because it was such a magnificent thing.

Anna Kenoff: I remember we called the engineers, so excited about this GaiaSoil that we found, and they were like, "No way, it does not weigh that much, that's less than water, it would float. You guys are crazy."

Patrick Hopple: We had been designing for a very normal-weight soil, typically 110 pounds per cubic foot. Well, Haviland was telling us, "We're using 100." I thought, Oh, well 100, 110—that's nothing to balk about. Then they came to a meeting and said, "No, it's 100 pounds total for each daisy!" I said, "That can't be right, you're messing up your units." Haviland said, "This stuff is really, really light." Then we looked on their website, and sure enough, this is one of the lightest soils in the world! And we had never heard of it, so when we got it, I was like, hallelujah, it works! It saved so much weight and money, we were able to reduce the number of bolts by nearly half. To put it in perspective, one of those columns was originally designed to hold something like six of Dan Wood's Mini Coopers. When we did the analysis again—with the lighter soil—we found out it was actually a lot lighter. It was actually only two Minis! The number of bolts went down by the same relationship. A weight was lifted. Everyone felt a lot better.

planting

Anna Kenoff: I grew up in Charlotte, North Carolina, the vision of suburbia. Charlotte is one of those new Southern cities that's just a tiny little downtown and tons of sprawling suburbs. My experience with farming: zero. I've really never grown anything. I didn't even have much interest in it. I think it's something that most people don't even think they're capable of, you know?

Sam Dufaux: Urban agriculture. I feel that there was a bit of a question mark there. We had people who told us we could grow things, but would it be successful? We were not too sure. The selection of plants had to be really resistant; they had to withstand loud music, spilled wine, the urban setting. Not necessarily the best environment for growing delicate plants.

Marcel Van Ooyen: I think they got a real taste of how difficult it is to be a farmer. I'd warned them, but they didn't believe me. I told them, "You're biting off more than you can chew because farming is hard, hard work." It's probably the hardest job in the world. And there is a huge learning curve. It seems very easy until you do it. Anyone who's tended a backyard garden knows the challenges that you face, from the environment to pests, fungus—anything! Finding the balance is tough. There's a lot of art to farming.

Haviland Argo: I made the first list of plants that I thought would work in the really dry conditions of that brutal, gravel-covered courtyard and elevated in the air. I grew up on a farm—cattle and tobacco—and every year we had a large garden. That helped, but I also ended up learning a lot from books that we purchased, like the grower's handbook—and then so much more from the people at the Horticultural Society and Michael Grady Robertson.

Michael Grady Robertson: I grew up in Kansas but in the suburbs, so I really didn't know anything about this world of farming until 2003. After graduating from college in 1998 and working some office jobs and traveling, I was looking around for something to do with my life. I started working on an organic vegetable farm in Texas, and once I started doing the work, I realized that that this was the life that I wanted, working with animals and the land.

Anna Kenoff: Michael has an interesting background. I think he was a philosophy major originally. When we met him, the Queens County Farm Museum had just hired him to enliven their program. It had been a farm for years but had become a kind of glorified petting zoo. I guess

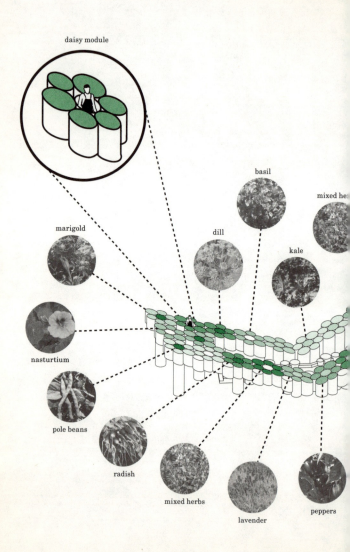

71 *planting*

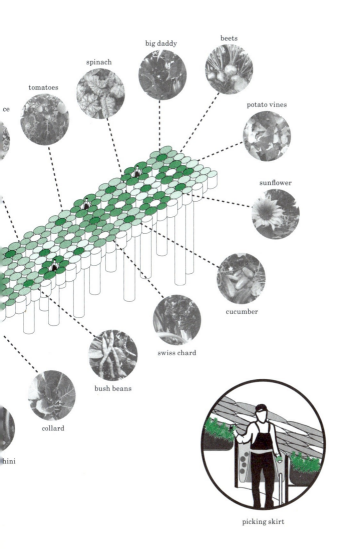

they saw an opportunity to get some young blood in to really make this into a working farm again.

Haviland Argo: Michael was really our go-to person, the person that was always there. He knew the most about getting started in the greenhouse, and he watched the plants, watered them every day. He reviewed every list we made for plant varieties and really believed in the project and was able to see that it could work. He was one of the first people to send us an email after the story in the *New York Times* broke, and we were making jokes that it was going be some old farmer in Queens, some hold-out with a little plot of land. It turns out it's this young guy interested in organic sustainable farming, working a fifty-acre farm!

Melani Pigat: Living in New York was an amazing opportunity for me, but I never in my wildest dreams thought that I would move to New York City and work on a farm.

Michael Grady Robertson: The architects came out to our farm and I think very few people had much experience in farming or propagating plants in a greenhouse. They would come out in teams of four or five, and I was able to teach them; they did an incredible job. They were all very helpful, even doing some volunteering for the farm. They would do a little bit of work on the P.F.1 project and then a little bit of work out in the fields here. We got them on the tractors, feeding the animals. It was a great relationship that developed.

Melani Pigat: The plants were an ongoing thing. Every couple of days and every weekend we'd go out to the farm and either plant new trays or repot trays that had already been started.

Mouna Andraos: I came down a couple of times to go to the farm and shoot videos of the farm animals. I was two feet in the mud with the pigs and goats, getting cameras in there and trying to get a better feeling for the farm.

The team was out there constantly planting, potting, and repotting in the greenhouse.

Sarah Carlisle: It was amazing to go out to this farm in New York City with skyscrapers in the distance and learn how to plant. It was something I never thought I'd be doing in an architecture job. When we went to the greenhouse and saw they had actually started to grow, we thought, well maybe this is going to work. Here they are!

building

Dan Wood: Things were in full swing: planting at Rikers and at the Queens farm, daisy fabrication and lacquering at the warehouse. There was so much more to do, though. We needed more time, more money, and more space; we were reaching our limits on everything. The only thing we knew we could get more and more of was people.

Anna Kenoff: It was a huge team. From WORKac there was Dan and Amale and Haviland and me. Sarah and Melani—from the beginning until the end—were incredibly important, making things happen on a daily basis. Melani had some construction experience and actually Sarah had a little bit of farming experience, so somehow together they were really a dynamic duo. Then there was Bryony on solar and electronics, and Magda in the warehouse. That was the core from the office, but more got involved as time went on. The larger team included people like Lenny and Marcel at CENYC, Michael, the Horticultural Society, GaiaSoil and LERA. These were the people that we were talking with every day. Then the volunteers, they just kept coming.

Art Domantay: In the beginning, I didn't want to rely on volunteers to help me out with this project. I didn't tell Dan that, but I didn't even want them in the warehouse, because I don't really work with volunteers. You know, they have their own lives, and I thought there's no way they can commit to this. But I was wrong. There were tons of volunteers that really helped. Without them, it couldn't have been done. They did a lot of the grunt work. As the builder, I had to figure out what job was good for what person. The carpenters were so specialized, we couldn't afford to have them just paint the Sonotubes. The Sonotubes needed three coats of polyurethane—that's thousands of square feet. The volunteers did that.

Sarah Carlisle: At first it was just Melani and me painting these tubes, three coats inside, outside, and on the edges of hundreds of tubes. It was incredibly overwhelming. Within a month, though, we had over fifty volunteers, and that also made us feel like, okay, this is possible. We have the hands, we have the help, and we have the young positive energy.

Anna Kenoff: The engineers also insisted we put this tape on the edges of all the tubes. We thought they were joking, but finally they said, "No seriously, you need to order one hundred rolls of this silver plumbing tape." Two layers of tape—top and bottom—to keep the tubes from sucking up water. Plus lacquer over it, which it turned out needed to be added all summer long, much to everyone's dismay.

Haviland Argo: We posted an ad on the Archinect website in the middle of May and were immediately flooded with people. We tried to get people who could work full time and not just weekends, but we had a policy of not saying no to anybody. It wasn't until we had a list of more than 100 volunteers that we started being selective.

Melani Pigat: Most of the volunteers who came on were students, just out of school. It would not have been possible without them. Sarah and I were not going to lacquer 360 tubes by ourselves.

Nicholas Muraglia: I grew up in Los Angeles, California, and went to Columbia University. Right after graduation they started construction on P.F.1, and I started volunteering. I've always been a big fan of WORKac. I studied them in school, their articles, projects. In the beginning I was lacquering the tubes and making them waterproof. That's how I started on the project.

Sarah Carlisle: We worked at the warehouse. The beginning of the day was usually pretty quiet. Mel and I would get there and the crew would have been there since 6 am. As the day went on, volunteers would trickle in. Some were punctual, arriving at 9 am; others came whenever they had time. We opened up the garage doors and did the painting outside. It was pretty much impossible on rainy days, but on sunny days we would roll the tubes out into the middle of the street and along the sidewalk. We would be painting hundreds of these tubes and cars would drive by and honk. People had no idea what was going on, but there were friendly neighbors who would stop by and bring us cookies for lunch. It was a fun time.

Art Domantay: The reason I wanted to be close to home was not just because I lived there but also because I really know that area, Greenpoint. The company across the street is a pallet-making company. Those guys are very neighborly; they took care of us. They helped offload truckloads of stuff with their forklifts and didn't ask anything in return. Sammy Lopez, one of the lead people there, literally saved this project thousands of dollars by taking whatever time he had to offload lumber, soil, pallets of bolts. He probably did that fifty times. It was exciting for the neighborhood, because not only did a lot of the neighbors get involved but they could see us working. We started running out of space fairly quickly and moved out to the driveway, then we were working on the sidewalk, and sometimes we needed to move right out into the street! I had to use one of the long Sonotubes as a barrier, so no one would get hit.

Anna Kenoff: There was a team in the office, coordinating volunteers and materials. We would order things in bulk, but in stages. I was on the phone with distributors constantly. We had this spreadsheet of materials twenty-three pages long. We would call in daily with these big orders, forty-six sheets of this kind of plywood and 1,400 screws and eight cases of Liquid Nails, always trying to get their pricing down. We tried countless people for bolt pricing. These are not normal bolts—they're huge. When they actually showed up, Art called and said, "Have you seen these things?" The wrench to turn them was three-feet long. There was a lot of calling the engineers to go out to visit and assure Art that this was in fact what they had in mind.

Patrick Hopple: Every weekend for a while there we were on-site; there was a large push to get everything completed. They would ask, "Is it okay if we do this?" We would sit down right there, do the calculations and

say, "If you're going to do that, let's do it this way." Just a real quick answer and they could keep working. They didn't have to wait for us.

Art Domantay: My job was to make sure I was feeding everyone materials, that there were enough Sonotubes and enough 2x4s and glue but also little things, enough blades and bits to go around. At the same time, money was coming in a little here, a little there. I was making daily runs to the lumberyard or Home Depot at 5 am, getting back to the warehouse and tidying things up. I was like Santa Claus, making sure everything was there for everyone before they got in. That was the rhythm, just trying to keep pace with these guys and making sure they had just enough to get by.

Sarah Carlisle: So much of architecture is not just about drawing. Anyone can sit there and, you know, draw up someone's details of a bathroom door or something, but to have the experience of working on-site, working with other people, with the construction crew, developing those relationships and trying to solve problems, communicating between different people and learning how to take on a leadership role was, I think, the most valuable experience that I could have as an architecture student. To take on the job of guiding the volunteers, not always knowing the right thing or knowing the answer but figuring it out or asking someone who did know—to develop that confidence was a big thing for me. I feel completely honored and very lucky to have had that experience.

Art Domantay: The people who put the daisies together stuck through the entire project. If one of those guys left, it would have taken a month to train someone new. For the people who were there from the beginning, it was more than just a job. I was always kidding around with Tim and Grady, two of the first guys, recalling the "good old days." When we were a month into the project, the "good old days" was like three weeks ago, with only two or three of us in the warehouse. It was different. It was quiet, and we had the whole space to ourselves. Four weeks into it, we had a bigger crew and you couldn't really tell them the whole history of the project. You'd tell them just enough to get started, and then away they would go. Their history was from that point onward. As people stuck around longer, the project meant a whole lot more to them.

Sarah Carlisle: My role changed and developed as I did. Once I gained some confidence and got to know the people I was working with a little better, I picked up certain areas that needed attention. Because it was just Melani and me on-site from WORKac, we had a lot of responsibilities and we had the independence to develop our positions. I know that I developed incredibly from the beginning. I was shy and a little reserved. I never would've imagined that I would ever drive a forklift, and yet there I was, driving the forklift!

Art Domantay: Everyone just got faster and faster. Every single crew member there, no matter where they started, in two or three days they became an expert at what they were doing. The warehouse would constantly change, morph into something else depending on the needs. One project would get done, tables would get moved, and a new station would get built. My office was eventually on wheels. I was getting pushed from one end to another every week.

Sarah Carlisle: There were so many assembly lines constantly going. There were six or more people cutting tubes at any given time, one or two labeling the 2x4s and blocks, a bunch drilling and fitting, the crew assembling daisies and columns, and so many people lacquering. When things were really busy, we would put more people on each station.

Art Domantay: Dan and Amale came to me and said, "We cannot be late on this. It just would be the death of the project." I never had a project that I didn't finish, and this wasn't going to be my first. For them and for me, it was about reputation. So we doubled up the crew during the day. We trained volunteers to do some of the precision cutting. I set up a nighttime crew, doing the initial rough cuts from 8 p.m. to 4 a.m. I had two teams putting the daisies together. We were maxed out.

83 *starting at P.S.1*

Sarah Carlisle: We kept going. Art and his crew just kept working away. Morale was boosted once they finished a handful of daisies; it seemed a little bit more possible.

starting at P.S.1

Art Domantay: I started going back and forth between the warehouse and the site. Sammy, me, and two other guys, we were loading and offloading the daisies and columns, using up the whole street. We had eighteen dollies that we utilized, because the daisies were too heavy to lift. We built a lifting device and used a forklift to carry these daisies on and off this rickety old trailer that Dan and Amale had borrowed from someone out in New Jersey to save money. It only fit four daisies, but as soon as we finished a trailer, I would call the trucker from the pallet company and he would truck it over to P.S.1. We needed to; we were running out of room in the warehouse.

Melani Pigat: More and more volunteers were coming every day. It gradually moved from doing the work ourselves to managing. Some days, we had thirty volunteers that we had to shuffle between the warehouse and P.S.1. Thirty people asking you, "What can I do? What can I do?" It was intense, especially on days when we were held up by factors that were out of our control. When there was not enough work to hand out, it was really difficult. We would try to do maintenance around the warehouse. A couple times, we sent them into the museum to take advantage of working at P.S.1. Usually, though, we just sent them back to lacquer the tubes, because you could never have too much lacquer on those things.

Sarah Carlisle: There were times when I just got very tired. It was hard work, long hours, with not much time off. Melani and I lived in Williamsburg, and we bought bikes to save money on the subway. There were some days when we finished up on-site at 7 pm, very tired, and had to bike ride home, which was a little intimidating and scary on those streets. Then we'd get home and have to carry our bikes up to the fifth floor, because the elevator door was locked after five. I would get home finally and say, I'm exhausted. I'm so tired. And we have to do this again tomorrow!

Art Domantay: When we began delivering the daisies, I started a second crew—the Install Crew—at P.S.1. They were the same four people I hired a few years before for a Mike Nelson project with Creative Time. That project was eighteen rooms plus eighty tons of sand piled on an armature that looked like a roller coaster. I had a lot of faith in these guys. But on the first day I told them about the project, they almost quit on me. They wanted to have a giant boom lift to connect the daisies to the columns. So I said we would try to have a boom lift, but we couldn't have it every day. Then Dan and Amale had to go find this boom lift, which was not in any budget.

Anna Kenoff: Art would call with lots of questions, like how do we get the assembled daisies off the jig? We need a gantry. Then he would just buy it. It wasn't in his budget, but there was no time. There was a lot of paperwork. Art was sending bills; we were buying things.

Art Domantay: Everything was so fast paced—no money and no time. Usually on a project, it's one or the other, but this one had neither. I knew I was costing the project money, but we had to act fast to get it done, and safety was a huge concern. Dan and Amale had agreed to this ramp. It really wasn't a ramp, actually, it was more like a framed, two-story house. I never said "ramp" to my crew because I didn't want it to be flimsy. I never even drew a plan for it. I knew they would make it work. These guys have all built houses. I figured, I'm not going to reinvent the wheel. They just had two days to come up with an idea and a week and a half to build it.

Melani Pigat: There were a couple days where it was just so hot—humid hot—and you could barely even move. Art's crew would set up a big fan in the courtyard and bring spray bottles with them, and everyone was misting each other in the face as they walked by. They'd spray water in front of the fan and make a big mist. Those times, everyone was moving a lot slower; it felt like any task was so huge. You were just so exhausted. You just wanted some air conditioning.

Art Domantay: When the daisies were trucked over to P.S.1 along with all of the columns, everything was laid out in the courtyards like a 3-D puzzle. Eventually, there was very little room left, so we were doing the same things over there that we were doing back in the warehouse— moving things around to make things fit. Every day we

would change a little bit, clear some space, because more stuff was continuously coming over.

Sarah Carlisle: At the same time, we were digging the column footings. It was hard digging. There was so much stuff in the ground from past P.S.1 projects: concrete, glass, metal. It took a few days with volunteers, pickaxes, and shovels. Even Dan dug one day. We had to go about a foot and a half deep and there were twenty columns, twenty holes.

Haviland Argo: It didn't really matter how much experience a volunteer had; what mattered was how interested they were in working and how strongly they felt about the project. Marcel at the Council on the Environment sent us this man; he had come to them and said, "For my seventieth birthday, I want to do a community service project instead of having a big party. Do you all have any that you're working on?" So we had about thirty-five people come out for his seventieth birthday and they just dug holes. We gave them lunch.

Sarah Carlisle: It was so exciting to see the columns go in. It happened so fast with the boom lift. Seeing them lifted that high and dropping them in—like it was nothing— was incredible. Someone would crawl through the column with a strap attached to the end of the lift, tie the strap around the column, and the crane would start to lift. One guy was like the cowboy; he used the rope to guide the

89 *starting at P.S.1*

column and walk it up the ramp. It would drop into the hole; then everyone at the base would prop it up and level it. When we had them all in, we noticed some were not level, so we all came around and twisted them: people holding the bottom, some up on the ramp holding the top, we would do a count "1...2...3...Push!" The bottom people would twist and the top people would push. It was a huge effort.

Melani Pigat: Eric's probably going to kill me for telling you this, but the second ramp that we built for the shorter end of the farm was a couple feet too high. He realized the problem before any of the daisies went up, which was lucky.

Art Domantay: Eric and Byron, both carpenters/installers, were staying with me and Marie. Not only was I working day and night, I had two crewmembers living with me. So that night over dinner Eric told me that he made this huge mistake. I kind of shrugged it off and said, what's for dinner? I knew the next day he was going to fix it.

Anna Kenoff: Just before they were starting to load the ramp with the daisies, the guys building it realized that the columns were not sticking out tall enough on the short ramp. It was a miscommunication, because we kept talking about how the ramp was mirrored but the short one actually started lower as well. That nuance never came through, and Eric, the main guy, just built it at the same slope as the opposite side. We had so much faith in Art and Eric at that point. Every time we checked something, it was perfect. We didn't notice it until they did. So after what I hear was a very sleepless night, they decided to all meet on the site at about six in the morning.

Art Domantay: The next day, he did fix it. All of the crew and the volunteers got in position. Two guys lifted up the entire ramp, balancing it on the two forklifts we had. Then everyone rushed in and cut off two feet of every single upright 2x6 with any cutting tool they could find. Everything got lowered back down in about two hours.

Melani Pigat: Fixing it actually just added more energy to the crew and the volunteers. It was exciting. We had no idea if it was going to work. So maybe it was a good thing in the end, because it got everyone really motivated and brought us all together.

Art Domantay: I had only one person that quit on me. I didn't say anything to Dan and Amale at the time, but that person quit because he didn't believe that we could do it, didn't believe that we could finish. He didn't want to be involved in a failed or losing project. So that says something about the rest of the crew, that even though we were up against the wall and stuck in a corner, everyone worked through it.

Dan Wood: Once the daisies started going in, we were into a whole new realm of equipment and expertise. I was signing checks and renting trucks, forklifts, boom lifts, gantries—a lot of this stuff; I still don't even know what it was! Lifting the daisies into place was a massive operation. Thank God we built that ramp; there were people up and down it all day, as the crew put the daisies in one by one, day by day.

starting at P.S.1

Melani Pigat: A few higher daisies were lifted with the boomlift, but for most of them we used the forklifts. We would attach a loop rope to the daisy so that the forklift could lift it up. A carpenter at the top would guide it. Then they would adjust it into place and take the planter disc out so they could get in there. There were holes already drilled into the daisy's side, but they had to drill through the column tops and then put the planting disks back on and insert the ⅞-inch bolts. At the very end, we fitted the planter shelves back in.

Sarah Carlisle: One moment I will remember forever is when we had all the tubes assembled and we had put the planter shelves in, but we hadn't put the plants in yet.

A bunch of us went up and sat in the tubes at the very top. We were having a barbecue that night, and we were just hanging out, thirty feet above the ground in these tubes. They fit a person so perfectly, just like little seats. Everyone was chatting and having a beer after a long day of work. It was really a beautiful thing to view, with the volunteers and the construction crew all sitting in these tubes, and the skyscrapers in the background at sunset on a nice summer evening. That was a moment of stepping back and really appreciating being there and all the hard work that everyone had put into the project.

Art Domantay: It was maybe the last week and a half when I started to let go a little bit. That's where the architects came in, and the volunteers shifted from just doing mundane things to doing the details. I told the crew, there's a certain point when the structure is up, the ramp is taken away, and now you have to give it over to the architects and artists. This is their project now. These guys, the carpenters, had the limelight through the whole project—up there, on the ramp, directing the massive columns. At the very end, even the fabricators had moved over with the installers; they were the kings of the hill. But when I knew our job was done, it was up to Dan and

Amale and their crew to do, not the finishing touches, but really, what everyone was expecting in the end. People weren't going to care about the ramp, how flat it was or if it held up. My job is done when our handiwork disappears and the structure is holding itself up. And the important part eventually was not the tubes, the columns, the structure inside there, the concrete, the brackets—that all became forgotten, and that's all good. What was important were all those additions that were placed inside and around the columns—the stories, the videos, and the plants. That was what made the piece to the viewer. What we have as builders, it's different; it disappears.

program columns

Nicholas Muraglia: It was very interesting to work with Amale's sister, Mouna. She's an artist who went to NYU's Interactive Telecommunications Program and was a fellow at Eyebeam, the art and technology center. Mouna was designing and programming all of the electronic elements for P.F.1.

Mouna Andraos: The idea for the program stations came from WORKac. They wanted to have little things happening in different places—benches and swings and plants, and maybe some sound and electronics. They knew that I could help with these experiences. Through dialogue and the exchange of ideas, we started defining all of the stations.

Nicholas Muraglia: Every column had its own specific event. It wasn't just an elevated farm; there was also a density of activity—electronics and other urban elements injected underneath for visitors to interact with, such as the Grove, the Kids' Zone, and the "Funderneath." WORKac is fond of those clever contractions. There was a juicer column with a table attached to it. There was a solar-powered phone charger, a sound column that randomly played animal sounds, a nighttime column with sparkly LED stars and crickets chirping, a video-iPod column, which had videos of farm animals; that was my favorite. All you would see from the outside is this column with holes in it. Once you look inside, it's a whole different world.

Mouna Andraos: I had to do most of the work from Montreal and tried to follow where the ideas were going and put different thoughts and tips on the table every once in a while. It wasn't necessarily super easy to not be there. I had a few people in Montreal helping me, but it was my first production project since moving here and we were still getting set up.

Amale Andraos: The program columns were really important to the whole idea that this was not just a farm but also an urban farm. The doubling of the space— with activity and party life below and the plants above— is essentially the kind of stacking needed to densify cities and introduce new programs.

Dan Wood: At the same time, we wanted to make sure it wasn't a purely pedantic project about plants and sustainability and all that. We wanted to make sure it

was a fun place to have a party. If people were interested, there were a lot of additional things to see and to learn; if they just wanted to sit down, have some privacy, get wet, or charge their cell phones, they could just do that.

Mouna Andraos: For months beforehand, they kept asking me for plans. I would say, I sent you the plan. That's as much planning as we do. I have to get started, and we'll figure out the plan once we're there. And so there was a kind of a cultural difference with the architects. Me, I have to build a prototype. I can't tell you how it's going to be before we have all of the parts and pieces. They were nervous about it, but in the end I think they were too busy to worry much. They had too much else to take care of.

Nicholas Muraglia: Mouna wasn't able to make any real plans, so I was put to work making construction drawings for the program columns. I didn't know CAD (computer-aided design) when I started, and it was a really good way to learn.

Mouna Andraos: The solar technology was its own adventure. We knew we didn't have the budget, so we needed to find donations and partners. I wanted to bring an engineer on board, because I didn't feel comfortable setting up the whole thing at that scale. So we contacted altPOWER.

David Gibbs: We are a solar-integration firm in Manhattan, specializing in integrative photovoltaics.

98 *P.S.1 becomes P.F.1*

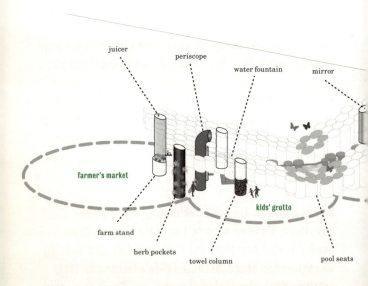

99 *program columns*

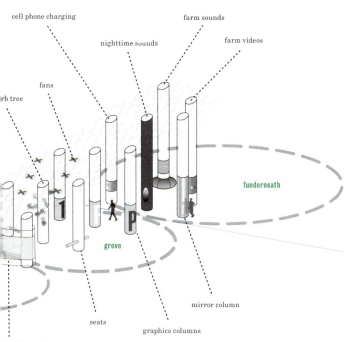

I became the head man because I had lived in Alaska for four years before coming to New York, doing off-grid renewable energy, and Public Farm 1 wanted to be an off-grid system.

Mouna Andraos: What I didn't realize before is that not many people know that much about off-grid systems. It's really kind of lucky that we came across David—even his own colleagues didn't necessarily know how to make these kinds of systems.

David Gibbs: Mouna does a lot of hacking, circuits, and chip programming. We began going over system components and determining load sizes. Mouna came down to our offices, and we started hammering things out. She had just gotten off a plane from Montreal and jumped right in, trying to get an understanding of how solar works. She had done small-scale stuff but nothing this large. It translates; you're just dealing with a few more factors: wiring, amperage, and circuit breakers. I showed her, "These are the pitfalls, and as long as you take care of those issues, you can do this."

Mouna Andraos: At first, there was some confusion as to whether or not they would be able to provide us with any equipment. Then we realized they didn't have any equipment, so we were like, okay, what good is this for? How is this being helpful? But then WORKac's team made a big push of phone calls and we looked through all of our contacts, and piece by piece were able to find these

donations. It was all quite improbable, because in that field everybody wants a freebie. We kept saying, what's the maximum we can get and we'll work with that."

David Gibbs: Bryony at WORKac was instrumental in getting a lot of the components, contacting a lot of the manufacturers. She also contacted a battery supplier out in Jersey, who provided all of the batteries that we needed. That was the meat and potatoes of it.

Mouna Andraos: So many phone calls. The pitch, of course, an "off-grid cardboard-tube farm for MoMA," helped a lot. It was good exposure for the companies. After every ten phone calls, one would have what we needed. Then, for example, we would get the solar panels and would need to get a truck and someone to pick them up in upstate New York. The batteries were donated, but they weigh a ton, and it was almost like a poison present. Nobody knows to this day what to do with them. I remember at one point, people saying, "well, worst case scenario we cannot plug it in, and we'll just have the panels there and pretend they're making power." I was like, no; we're beyond this. We need to make it happen for real. Until two or three weeks before, we didn't have all the pieces.

David Gibbs: They knew what they wanted to run—the lights, the pump, some fans, smaller electronic devices. You figure out how many hours you want to run these things and how much energy you're going to be burning. You have your load side, then your battery side, which

102 *P.S.1 becomes P.F.1*

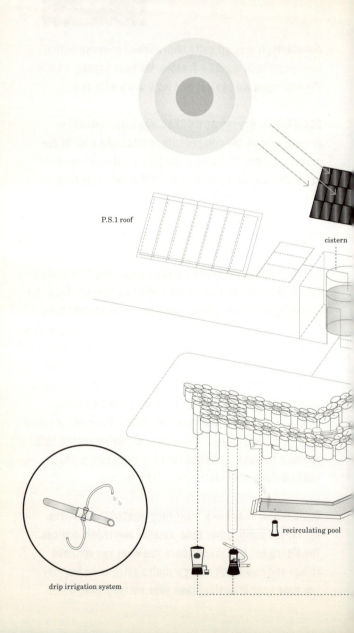

103 *program columns*

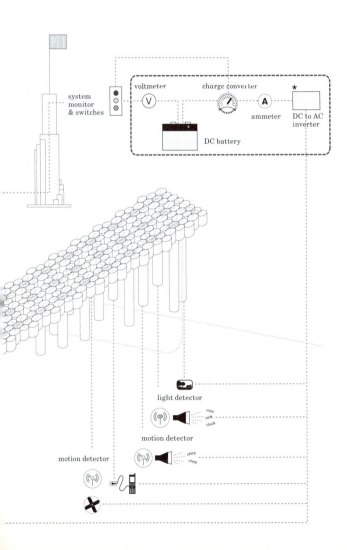

is essentially like your gas tank, and then your supply side, the photovoltaic modules. You have a set amount of sunlight: an average of about six hours in the summertime. From that, you can determine how fast the energy is coming in. You already have how fast it's going out, and then you can size your storage. There are other smaller determining factors too, but that's the general idea.

Mouna Andraos: We had so many discussions about where to put the gear. We knew we needed a storage shed for the equipment, and there were a lot of "form vs. function" battles as to what this was going to be, whether it was going to be safe and look good, who would have access to it, and how it would be maintained. In theory, I think we should have had someone who was able to maintain it much more thoroughly, but we got away with not much maintenance over the three or four months. They eventually put together that tower mostly for form, more than for practicality, but it worked to hold the equipment.

Sam Dufaux: I was a little concerned about the side things. The structure is one element, but then we decided to add the Tatlin tower with the solar batteries; we had the solar panels, and we added a cistern to collect water. At some point I thought, there's just so much, and I'm not sure you literally want to build an entire farm. The structure is enough; we should put all of our energy into it. But then we found good ways to integrate everything. All these things that I thought were superfluous, at the end, were really done to widen the audience. In previous years, there was

105 *plants on-site*

a sensation that the installation was built for architects, but here you could have families, you could have kids going to a party, you could have nature aficionados. All these little parts really made sense at the end; everybody found a piece of the project that they could relate to.

Anna Kenoff: For me, one of the important things was watching Dan and Amale—the way that they operated and their relentless insistence on not letting any parts of the original design get washed out as we ran into budget and construction realities. There were times when I would question things like, we're in this major crisis, is this little tiny piece so important? Dan would always say, "Never kill ideas." In the end, those little things that they insisted on keeping were what helped bring the loftier ideas to life. The ability for someone to interact with and charge their cell phone with solar power is a seemingly small detail in the overall project, but it was really important in terms of the way people could grasp the wider concepts.

plants on-site

Marcel Van Ooyen: We figured, if they're building this, they probably wouldn't want to use a lot of municipal water, and we might as well try to capture it from the buildings around there. So we suggested a water-collection system. We've built about thirty of these around the city and actually won an innovation award for them two years ago. I always laugh at the idea that capturing rainwater is now an innovation!

Lenny Librizzi: I work for the Open Space Greening program at the Council on the Environment, assisting community gardens throughout New York. Starting around 2001, we began to install rainwater-harvesting systems, and we told WORKac that we could help put one at P.S.1. We borrowed a tank, they found a company to donate the drip-irrigation system, and they were both hooked up to a pump. Then they used solar power to power the pump. It was a pretty impressive system.

Paul Mankiewicz: Now we had to find a way to hold all the plants and soil within the tubes. WORKac found these very neat baglike structures called Smart Pots that could sit on the plywood shelves in the tubes and hold in all of the nutrients while draining the excess water out.

Amale Andraos: The Smart Pot company donated all of the pots and made them in custom sizes to fit the three diameters of our tubes. They were a great find, because we were always worried that mud and dirt would be dripping on people after a rain or after irrigation. The Smart Pots were a kind of filter that made sure the only thing dripping through was clean water.

Anna Kenoff: We started the seeds in February and March, and some of the seeds grew faster than we expected. As we gradually moved them to the site in flats, we repotted them with the GaiaSoil, which requires all of these different layers; so the process was a huge, ongoing effort. There was a team constantly potting, handling, and organizing the plants, right up until the opening.

Sarah Carlisle: I taught the volunteers how to repot. There were thousands of plants, so it was a pretty big job to make sure they were all in the pots the correct way. We filled the Smart Pots with GaiaSoil eleven inches deep. To do that, we lifted these huge bags with the forklift, then opened the bottom, and released just enough soil into the pot before closing it off, like a grain elevator. We put a layer of jute over that, and then about an inch of compost on top.

Paul Mankiewicz: The GaiaSoil is not a surface soil; you can't just pour it in and plant in it. You have to recreate another feature of soils, the O Horizon, which is the top layer of soil. If you take a spade to a meadow and dig down, the top layer is humus, which is composed of organic matter and mulch, dead branches, and leaves. Below that is the topsoil, where it's all mixed in and where the roots mainly grow. GaiaSoil provides that lower zone for the growth of plants and roots, holding everything together. By putting a layer of jute over the Styrofoam, we're essentially holding it in place so it can't blow away. Then we put compost on top of that—as the higher nutrient zone, where the plants pick up much of their mineral requirements. They got all the compost from the wonderful program—New York City Compost Program—at the Department of Sanitation. They will give any not-for-profit company all the free compost you can haul.

Sarah Carlisle: I went on the first run with a volunteer on a truck over to the Fresh Kills Landfill. We each had a shovel, and planned to put the compost into these huge bags that are about five feet in diameter when full. We went out there and encountered this huge pile of manure. We shoveled the bags, maybe only a tenth full; they were so heavy. We had to make quite a few trips.

Melani Pigat: There were some days when we would come in in the morning and the plants did not look so good. That was a big fear. If we had this amazing structure all set up and ready to go, and there were no plants, it

would have been a total failure. They were like our babies. We had no water pressure at the beginning, so we would spend hours in the morning watering; it took three or four hours, by the time you got to the last row of plants, which was way later in the day than it should have been. Then at the end of the day, we'd stay and water the plants again. Eventually, we hooked the hose to the fire hydrant outside.

Michael Grady Robertson: It was a little tough for them to water things. They didn't have enough water pressure at one point. I think that might have actually contributed to the well-being of the plants—a little bit of stress. It's not like you always want everything to go perfectly. Sometimes in order to overcome and to achieve, even plants need a little bit of duress. So I think there was some concern about the watering, but it wasn't a disaster.

Anna Kenoff: I remember we all just kept being terrified about how the plants would fare. Michael taught us a lot. He would come over to P.S.1 after work, at 7 p.m. on a Friday, when we would call him because we were freaking out that maybe one plant was wilting. He kept saying, "They're plants; they just want to grow. Let them grow."

Sarah Carlisle: We had determined with Michael how many plants could go in each tube depending on each plant's specific needs. The plants were all in the back courtyard at first. We wrote on pieces of duct tape on the walls of the courtyard—the number and letter of each daisy, the types of plants to go in it, and the number of plants per plot. The tape was placed to correspond with a row of pots laid out in front of it. We started with plants that were overgrowing their pots, so unfortunately they weren't organized according to their placement within the structure. That made things a little confusing, but the duct tape labels worked out pretty well. Soon we ran out of space and had to move out into the larger courtyard, where it was harder to find enough space. There were little strips of duct tape everywhere. It was right up until the end, but we eventually had plants left over, which we gave to people in the museum.

Anna Kenoff: There was all this build-up on-site, to the day we were going to put in the plants. People were excited that they could stop spending four hours a day watering. We felt like we were behind—that it was one more day, one more day—but we had to wait until the irrigation was set up and working.

Lenny Librizzi: We were collecting water from the roof of this loading dock, located on the other side of a concrete

wall, from the cistern. The museum wouldn't let us punch any holes in that wall. Haviland said, "You can't go over the wall. You can't go under it. You can't go around it." Luckily the wall had these very small, one-inch deep holes, and we used those to run small tubes from one side to the other. So water collected on one side in a four-inch pipe, had to go through this series of smaller, three-quarter-inch pipes, and collected again in another bigger pipe that funneled it into the tank.

Nicholas Muraglia: There were some stressful moments. At one point, there was a forecast of hail, which was very strange. It was the beginning of summer. And so there was a huge panic, because all of the plants were just laid out in the courtyard, exposed. The clouds were coming, and it was really dramatic, dark, and ominous.

Lenny Librizzi: In New York, even during the hottest months of the year, the city gets a significant amount of rainfall, practically the same amount as we get in the spring. In August we get more than we get in April, which you wouldn't think makes sense. The problem is that it will come in these downpours, and then you won't get rain for a week or two.

Anna Kenoff: Amale was very protective of those plants, as she should have been. They were so beautiful when they were on the ground. One Sunday we were all working and there was supposed to be golf ball-sized hail, and so everyone just panicked. Amale started ordering everyone into gear, finding anything they could possibly find that would act as a tarp and covering up all of the plants. We didn't want to crush the plants, so we had to build these tent structures, which, of course, blew over in no time. And so ultimately we ended up with piles of tarps, uncovered plants, and we were all just sitting on the steps of P.S.1, drenched and pouting, waiting for the hail to come.

Lenny Librizzi: The day we had the cistern hooked up, we got a torrential downpour before we even really finished. Water was backing up all over, because we had run forty feet of pipe behind the wall to the downspout from the roof, all the way on the opposite side from where it needed to be. The water was bubbling out, right where it was coming off the roof. So we had to immediately go back—in the rain—to make an overflow piece.

Nicholas Muraglia: It was a huge rush to get tarps to cover all of the plants—super difficult. Then they all blew away and there was nothing to do. So we left. In the end, it was funny how dramatic it was because, you know, I guess plants actually like water! And there was no hail.

113   *plants on-site*

Patrick Hopple: In the computer model that we created, we didn't really know what values to use for paper's material properties, so we estimated very conservatively just in case the paper got very wet by the end of the summer. There was this huge rainstorm during the install, and they called us; while we were secretly praying to ourselves, we were saying to them, "I think it will be okay."

Amale Andraos: So that first storm was a good test. I was super nervous, but the plants all survived and we calmed down a bit after that, and, of course, we now knew that the paper could stand up to a huge amount of wind and water. Afterwards, there was a lot of re-lacquering and re-taping everywhere we saw stains.

Anna Kenoff: Finally the day arrived—a couple of days before the opening—to put the plants in. We got everyone up on the short side of the structure, with an assembly line from the ground to the top. We formed a chain and began passing the pots one by one by one. It took four people to hold one pot.

Sarah Carlisle: We weren't sure how long the process would take. I had a diagram of where all the daisies were in the courtyard, and it was a lot of work to make sure the right plants were going in the right tubes. The top of the structure was difficult to reach, so we used the big forklift rigged with a homemade pallet that could hold six pots at a time for the people on the top.

Anna Kenoff: Sarah had written in pencil on each planting disk which plant went into which daisy. We'd stand up there and call down to the team on the ground, "We need a thirty-inch beet! Get me collards!"

Sarah Carlisle: Emotionally, I was pretty stressed out. I wanted to be a part of it and wanted to help move the pots, but every time I got in there to give a hand and help with the work, someone would start yelling that they needed some bell peppers or something. I was the only one who knew where all the pots were located, so I would have to run around and show everyone where they were and what to do. It was a struggle for me to be in charge, a sort of character struggle.

Dan Wood: Some of the pots went in fine. Most of them were packed too full, so it was difficult to get them into the tubes. Once the pots came in, we had to take the coiled-up irrigation pipe out of the tube and drop the pot in. Then everyone had to get their hands between the pot and the tube, and push as hard as we could to get it down to the shelf. For weeks afterwards I looked

like a street fighter with these huge raw marks on my knuckles.

Anna Kenoff: It took a long time, the process of finding the right plants, getting them on the forklift, getting them up. We started a second line, working from the bottom up. Sarah took over the small forklift—she was a beast with that forklift—and organized the second group so we could meet in the middle. It was hard to get down from up there, but we figured it out. It went pretty quickly. We had pushed the day so close to the opening, but once we got going, we knew we were going to make it.

pre-opening crunch

Mouna Andraos: When I finally got to New York for the install, two of my friends jumped on board and helped out. That was a lot of fun because I had been alone in my studio, doing all that work and just talking on the phone. Now I finally got to be on the ground. The energy of the site was really great.

Andres Lepik: In the beginning there was a small group of volunteers, and then, whenever I visited them,

pre-opening crunch

there were more and more people that we hadn't seen before. When they built it up in the courtyard, there were like really, I don't know, a hundred or so. It was really incredible.

Haviland Argo: We provided lunches for everybody; even just making the estimate before everybody arrived—to place the order for those lunches—was a ton of work.

Mouna Andraos: Everything I wanted to do involved inserting things into columns, and there was a lot of pressure from the engineers to make sure the structure would be okay. Amale and Dan were like, "Enough with your holes. All you want to do is make holes in our columns!" I said, "Well, I need to get in there, one way or another." So we spent a few days negotiating. We were looking at this huge structure, saying, "It's not going to fall apart, you don't have to worry."

Anna Kenoff: The whole thing was really fast and furious, but it also evolved a lot. The ideas stayed the same, but the design was ongoing until the last day.
Mouna Andraos: I made a disk with all of my little

constellations for the night stars column. Then, of course, Dan sees it and says, "This is way too big. You're never going in with this." I was like, "What? I worked on this for days!" He said, "Well, you'll have to cut it in two or find a way to fit it." I spent two days figuring that one out.

Nicholas Muraglia: We were working with these very sensitive electronic interfaces, and we had to deal with all of these water issues. It's an outdoor project and had to last for months in the rain. The video players, for example, had to be on constantly; we had to devise all of these interesting ways to protect them. We ended up encasing everything in Tupperware.

Melani Pigat: There was a lot going on. Toward the end, we were installing and making finishing touches to the programs components of the project. The midnight column had been painted but wasn't quite the right color,

so I was repainting it. Late into the night—any time after 9 p.m.—we were working in the dark. I couldn't even see what I was painting!

Mouna Andraos: We were often on the late shift. I might have been a bit more on a tight schedule at that point, closing the place every night.

Anna Kenoff: I remember starting earlier and earlier. The graphics crew was there, hanging the columns and painting the logos.

Kevin Wade Shaw: I was an intern at Project Projects, the graphic design studio that shares space with WORKac. I had to paint these giant letters that said "P.F.1" onto the columns. Amale had a technique from a professional exhibition painter; it was very exacting. We printed these giant letters onto paper and then hung those on the columns. I spent all day arranging them, making sure that when you looked at them from a certain distance, they lined up. Then it started to rain and the ink on the paper bled everywhere; two out of the three letters fell off. That was fun.

Elodie Blanchard: I did the fabric for the columns. They were held on with Velcro, grommets, and screws. They were just the decoration on top, like clothes. They gave me a technical drawing—drawings from Mouna, showing what she needed—and from there I figured it all out. It's all outdoor fabric. They wanted something more natural, but the cardboard is already beige and natural looking; I said it would be nicer if the columns were more colorful and funky. It gives a nice contrast between the organic feeling of the plants and the cardboard, then all this color.

Sarah Carlisle: We had to fix the tape on the underside of the tubes that had gotten scuffed up during construction. This was at one of the highest points, and you had to stand at the very top of a ladder to reach the bottom edge. Fixing the tape was one of the volunteer's jobs, but none of them felt comfortable going up there. So I was like, okay,

pre-opening crunch

I can do it. Art gave me a harness, a huge massive thing, and I'm a relatively small person, so I had to cinch it up quite a bit. They hooked me up, someone held the ladder, and I climbed up and fixed those edges.

Anna Kenoff: At that time, Art was trying to not be involved, and failing. Everyone would run to him with questions, and he had a lot of answers. His crew was there, more in the background; they sort of redistributed themselves when we needed a more expert set of hands. We had a full-time crew taping and lacquering. It would rain, we'd see spots of water, and hit those again. A team in the office was laser-cutting stencils with the names of all the plants so that they could be identified from below. And there was a team on-site spray-painting those names on the undersides of the discs.

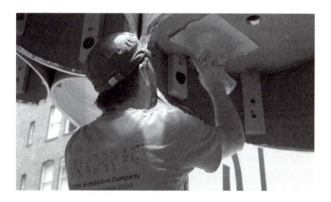

Elodie Blanchard: Everything changed all the time. They had an idea with the bench, for example, then it changed once they started building it. They were like, "Oh yeah, it was like that but, in fact, now it's different." I gave some bad ideas too. They made this nice tube as a seat with black rubber and I said, maybe you should put some color fabric on it. Then they were so ugly. They put color fabric on all the seats and then took it all off again.

Sarah Carlisle: There were volunteers cutting holes in the columns and inserting the bench seats—the smaller diameter tubes—testing how to cut them.

Elodie Blanchard: It was all so last minute. For me, it was crazy because I had a one-week-old child! There were a lot of things I didn't think worked really well, but I didn't have time to order all the fabric and a lot ended up not being UV-resistant. If I had done a little bit more thinking, and had more time, it would have been better. It was like, Oh, I have fabric in the basement. Gee, it's green. I'll just take that! My baby had just been born; my mom was here. I would have loved everything to be finished before, but in fact nothing was finished before.

123 *pre-opening crunch*

Anna Kenoff: There was a whole group building the pool—a much slower process than we anticipated. The foam was more difficult to work with than expected, and we were reusing plywood that we didn't want to throw away. Then the guy came to spray it. He started but then went to move his car and was gone for six hours. He got in some kind of accident.

Kevin Wade Shaw: We had painted these letters on without any sort of a mock-up of the vinyl posters that would go on the opposite side. There was this really hectic moment when we were sitting in the studio, and I got a call from Dan. They were trying to hang the vinyl banners, and I didn't know exactly how to tell Dan how to hang them. If you're looking at the columns from the front, you're not supposed to see the banners at all. Everyone at the site was trying to figure out how to put these up.

Anna Kenoff: The last few days are such a blur. Art kept trying to make Sarah, Melani, Haviland, and me carry around clipboards so we would focus on supervising instead of working. Art has experience running projects like this; he knows how important it is to keep some people focused on the overview, not the details. I was checking in on the irrigation, the solar installation, this whole team running all the wires, burying cords, making sure it worked, connecting everything. Art thought the clipboards would help, but we kept losing them.

David Gibbs: We couldn't run a bunch of hard conduit to create runs of wiring to an outlet, so we just made runs of extension cords with a surge protector so we'd have multiple plugs. Even when we got it all together, I remember we turned it on and something wasn't working; we weren't getting the voltage readings. I was getting panicky. Then I realized that I didn't hit one of the other breaker switches, so it wasn't getting full voltage!

Anna Kenoff: There was another team working on the Tatlin tower. The base was already there for the solar install, but the rest had to be built and put into place.

David Gibbs: Doing the wiring of the array was really difficult. Mouna, Bryony, and I were doing that for long

hours on the Genie Lift, and she'll probably tell you, this was not fun. At twenty feet in the air, the only way to get around it would've added a lot more expense. Because labor was free, we put more time in.

Elodie Blanchard: Also, the yellow fabric at the bench was all wrong. It was really heavy and too stiff in the middle. It was an egg-yellow color and just didn't work with the beige. Amale sent me a picture, and I was like, yeah, that looks really bad. It was the day before the opening, but we had to change it. So I just put something brighter; the new color worked better. The lighter fabric moved with the wind and kind of flowed—a touch of yellow in the middle of the structure.

Kevin Wade Shaw: Amale said that she would like Project Projects to put together some printed materials for P.F.1. This was the day before the opening. For a spread in *Time Out*, we had taken a text directly from *Ecotopia*, a sci-fi novel about a non-techno future; so we built on this idea. We pulled all of these texts and images from

Project Projects's and WORKac's libraries—books about architecture, agrarian lifestyles, DIY cultures, radical architecture groups of the '60s and '70s, etc. We looked through tons of books, scanned them, and then laid out a seventy-two-page booklet. Adam and I finally finished the design by four in the morning. Then we had to find a print shop. We eventually found a 24-hour Kinko's on the Upper East Side, a real trek. We sent the files out at 4 a.m. and we both went home. I got maybe a half-hour of sleep. Kinko's apparently couldn't staple and fold at the same time, and when I got there, they were making them all backwards. I had to show them how to fold and staple. Finally, we got it all worked out. I got the booklets and made it back to the studio just in time for everyone to go over to P.F.1 for the opening. It actually turned out to be one of the more interesting components of the entire graphic project.

Adam Michaels: I've collected 1960s paperbacks for years, focusing on those with complex relationships between texts and images—Marshall McLuhan and Quentin Fiore's *The Medium is the Massage* is the classic example. I keep a stack of these next to my desk at the studio. The day before the opening of P.F.1, Amale came by and asked if I would lend some paperbacks for placement in the installation. While I try not to be too precious about these things, I wasn't really into placing a segment of my book collection outdoors in Queens. My counterproposal was that we make a Xeroxed zine to be distributed at the opening, which would be happening just thirty hours after

our conversation. So this was a completely crazy process, but I'm still fond of the outcome, and that activity was definitely important for getting things rolling toward the Inventory Books P.F.1 paperback.

Sarah Carlisle: There was just so much going on. There was a periscope that one of the volunteers had made at his father's sheet-metal shop. That install was pretty last minute; final touches were happening on the morning of the opening.

Anna Kenoff: The flag! Dan climbed up on a pallet on the forklift—got hoisted up with the harness—to plant the flag on top of the Tatlin tower. And we had the graffiti artist, Diva, spraying one of the extra tubes to make a big graffiti bench. Up until the last minute, we were filling the holes around the columns with concrete to support them.

Elodie Blanchard: The engineer was so stressed out. He was like, "Can I put a piece of metal here? This is going to fall." I mean, these tubes full of dirt are up there in the air.

Patrick Hopple: The project is sloped in such a way that it could actually catch wind from both ends, like a sail, and it would want to move. There was some back and forth about solutions. I had to run out there, basically at 10 a.m. before the opening. I was in a nice suit; it was 95 degrees and I was up on the forklift, taking some pictures, doing some measurements, making calculations, and Dan was like, "That's what I'm talking about—engineers doing engineering work!"

Dan Wood: I remember one of the forklifts got stuck in the small courtyard and we had to push it out of the gravel. Then, I turned around, and the site was spotless. People had left to change for the opening, the caterers were setting up, and the farm was this really brilliant cacophony of greens in the air. The sun was catching the silver plumbing tape and the structure looked massive and solid, with an incredible variety of colors and environments underneath. I was like, where did this come from?

first impressions

Patrick Hopple: I was really happy seeing other people interact with it. The day it opened, people were out there touching it, knocking on it, saying, "It's paper!" I brought a few of my professors from college, and they said, "This is amazing. No one would ever do this." And I said, of course we would!

Kate Chura: I would stop in as much as I could, and you would hear people talking: "What is that? What is that thing growing up there?" Inevitably somebody else walking around would know, and this whole dialogue would break out about what you can grow in the city. We all pick up information and process it in different ways, but for many of us it's physical and visual. P.F.1 just verbally opened that up for people as they walked in and moved around it.

Elodie Blanchard: I like Dan and Amale's design, their aesthetic, their way of doing things. You know, you can complain that they're crazy—that it's all the time last minute—but I don't care. I like that they have an idea and they're never going to say, "Oh, it's easier like this," and make it less strong. They have really strong ideas, and they will do everything to get there. I know I'm not like that; I really admire that they can have a concept and just go to the end, and put all that energy in. You have to be really strong minded and confident to do that.

David Gibbs: I was blown away with how big the structure was. I mean, it was the size of a house. I was really amazed.

And I remember Dan said, "Yeah, it was supposed to be bigger." I was like, really? Okay. Very ambitious.

Nicholas Muraglia: From outside P.S.1 all you could see was a hint of the farm just reaching over the concrete wall. Once you went around the wall, then you were just hit in the face with this huge, crazy structure. When I was working near the entrance, I would see kids and parents coming in with their mouths wide open. Just like, "Wow!"

Michael Grady Robertson: That image of this incredible cornucopia within the structure was really amazing— its sense of abundance, up against the concrete slabs and coming up over the wall, with the plants hanging down. I was standing inside, looking up at the sky, and looking at the spray-painted building across the street and thinking to myself, this is such a novel perspective.

Paul Mankiewicz: I was absolutely floored. The fact that they were using these humble molds for pouring concrete in those rosette shapes; I thought it was multiply symbolic— taking something that is normally a part of cities and infrastructure, and turning it around to make an elevated farm. Also, cardboard is basically a laminar cellulose. As a biologist, I deeply appreciated that because the cell walls of plants also have a laminar-cellulose skin that is very strong in tension. It was just a beautiful analog to the way nature works, having these extremely strong tension-held, low-cost cylinders rosetted together and filled with soil to create a loft of light-gathering edibles and flowers. It had been a long time since Queens was quite so well inhabited.

Dan Sesil: The idea of using materials that were recyclable was very satisfying; that we could pick a material that would not impose itself on the environment in a bad way afterwards was really cool. Also, you knew immediately when you saw it, the various levels that the project worked on. It provided food. It was sustainable. It harvested the energy of the sun and water from the area. And it was a very recyclable structure. There was a message there that wasn't heavy-handed; it was really easy to understand.

chickens

Marcel Van Ooyen: Chickens are mean. They smell and they require a lot of care, but they wanted them.

Michael Grady Robertson: We have a program where schools will hatch out fertilized eggs and return them to the farm, where we raise them as laying hens. So we had some fairly young chicks, and we thought, well, animals are a pretty important part of any sort of farm project. Plus, kids and families really seem to take to livestock and animals. I think that maybe we took a bit of a chance with that in terms of having a real chicken coop on the museum's grounds, but it seemed to work out okay.

Lenny Librizzi: Initially, I was not sure that it was going to be a good match, a water tank and chickens, but it seemed like it worked. They lived happily together for the months that they were there.

Mouna Andraos: Where to put the solar panels evolved from the architects' vision of round panels on top of the structure, to the reality of having to deal with these much bigger, not as pretty, rectangular solar panels. That was definitely WORKac's touch—to make them into a rooftop for the chicken coop.

Dan Wood: The whole time, we were telling P.S.1 that we were building a tool shed to hold all the tools in that small courtyard. So we made a door, put a lock on it, put the cistern in there, and covered it with the solar panels.

Amale Andraos: The day before, we made the roosts out of extra tubes, with little ladders leading up to them, and brought in a bunch of hay; about an hour before the opening, Michael Grady Robertson came in a white suit, carrying a cage with five big chickens and a box with twenty adorable little chicks peeping away.

Andres Lepik: They kept it a secret that they would have these live chickens there. I think only one or two people at P.S.1 really knew about it, but they didn't tell anyone. When we came for the preview and there were the chickens, this was a really great surprise. And it was also fun during the opening, when the chickens were running around. This was really the greatest surprise of the whole project.

use

Haviland Argo: I really liked seeing the kids that came to P.S.1. All along, you think that you're designing for the adults who are coming to drink beer, listen to music and party, but the place became very popular with the under-five crowd. They would show up, take off all their clothes or strip down to their diapers, and play in the pool. We also learned very quickly that the kids don't stick to prescribed ideas about how you act around certain things. They would use the rubber seats that we made as trampolines and just jump up and down on them. They were really interested in how things were put together. They wanted to rip the fabric off of the columns and push the animals to see if that's how they made the sounds.

Mouna Andraos: The day after the opening, I was standing there and two families showed up with a bunch of little kids. They undressed them, threw them in the swimming pool, let them run around, and in half an hour these kids were all over the place—jumping on the seats and pulling on the fabrics. They were jumping on the inner tube, holding themselves on the fabric, pulling as hard as they could, punching in the speakers as they were jumping, pulling on the phone strings. We were just sitting there, thinking, there's no use in stopping them because tomorrow we won't be here; so we better just see how far it's going to go. I think they did all the damage that was going to happen for the entire summer, all in that first hour. After I left, WORKac added more protection to everything.

Anna Kenoff: I think we thought it would get a lot more trashed than it did. People seemed to have some respect for it. I mean, we put up a lot of "do not climb" signs to keep people down, but we really didn't have that problem.

Kate Chura: Daily maintenance on the plants was handled by two members of the GreenTeam, people who have gone through our Rikers program and had been released. It's an entrepreneurial program where we go out and bid on landscaping contracts, and use those to keep them employed. That's the most vulnerable time of their life, and the ultimate goal is getting somebody stable and making sure that they don't end up back at Rikers.

Michael Grady Robertson: The GreenTeam really became quite fond of working there. They became the caretakers and the mothers of those young chicks.

Anna Kenoff: We had rigged up this fantastic irrigation system, and no one told us that there was a filter in the pump that had to be cleaned every now and then. So, of course, our cistern is taking water off the roof, which is full of all kinds of smut; it clogged in the first week. The first time the GreenTeam went out there, they called the office and said, "Oh my God, all the plants are dying," and I remember saying, really? All the plants? Sure enough, they really were wilting fast. So we got this guy out to troubleshoot, and found out it was just the clogged filter. Then, once we put water to them, the flowers just perked back up. I really was afraid that we were going to have to replant everything, but even the sunflowers came back around. It worked out.

Nicholas Muraglia: There were many times I had to go back out to P.S.1 just to help keep the farm going. We had these fans attached to the underside of the structure, and one of them broke at some point, so I had to go fix it. It was late at night, and there was no one there—just me standing on this ladder, very high up in the air, fixing the fan. I got it to work and was watching it run, just to make sure. All of a sudden, this huge deluge fell on me. I had totally forgotten about the timing of the irrigation system! And when that irrigation goes, it really runs—like a typhoon! This was during the summer; so at least it was warm, but then I remembered that the night before, we had put this kind of compost tea, called Terracycle, in the cistern. So it's the middle of the night, I'm standing underneath a cardboard farm, fixing a fan, and now I'm covered in worm poop! I don't think many people can claim to have experienced that. We had a time-lapse camera, so I actually managed to find that exact moment later—you can see me dripping wet in the middle of the night.

Elodie Blanchard: One of the problems that happened is that I took some fabrics that weren't UV-stable. I thought to myself, two months is fine, but two months was not fine. It was really ugly at the end. I said, oh my God, did I do that? It was bad. I said I should do something to fix it, and then they were like, "It's okay. We took pictures." But it looked bad at the end. So bad. Next time I know. I learn on every job.

Mouna Andraos: We were worried a few times; every time I got a phone call from WORKac, my heart stopped beating for a second. Uh oh, what have I done? Sonali, who was in New York, was on call. She knew how everything had been set up. When she went to do one repair, she said that one of the electrical boxes was filled with water. She was completely freaked out when she opened it up, and we still don't really know what happened in there.

Anna Kenoff: I'm a little biased, but I got a lot of tremendous feedback. We had to go every Saturday and Sunday to water and take care of the plants because the GreenTeam worked only Monday through Friday. It would start to get crowded pretty early, and you couldn't even get the work done without people wanting to know more about the farm, how it works, and what you're doing. We had people calling the office all summer, asking questions.

Dan Wood: Harvesting was great. We would get a group from the office, bring the skirts, and go up through the picking holes on ladders; later we learned how to walk up and down the structure, balancing on the rims of the tubes and then sitting in the picking holes to weed, pick, or snack.

Amale Andraos: A lot of the produce went to P.S.1. They used it in the cafe or brought it home. Sharon and Jamal, the GreenTeam farmers, also brought some home. We took some to the office. Once the chicks grew up and started laying, the cafe started making a lot of egg dishes.

Barry Bergdoll: Since it was a freestanding object in the courtyard, rather than something that really used the whole space, I was afraid that the partygoers would not adopt it; they would think it was something they needed to get under on the way to the party. But they had all those fun things in the columns and, from all the reports and videos, it was incredibly, incredibly popular. I received a number of written notes and anecdotal things—people thought that it was the most fun and the best ever. That people said it was fun—that was a great relief.

Sarah Carlisle: We went to the first big Warm Up party, and that was incredible. Just seeing that many people in the space, and walking by and hearing other peoples' opinions on something that we'd been working on for months. The place was packed, and people were so interested in the project, and really interacting with it. To be able to see a project through from the very beginning in such a short amount of time, that's an experience that most architects don't ever get.

Melani Pigat: We got to go to one of the Warm Ups before we left New York. It was really rewarding, going and seeing thousands of people enjoying it. You could tell everyone really liked what we had done. The energy was amazing. I had never experienced any kind of party like that in my life, so that was pretty cool.

Dan Wood: I think it was even in the *New York Times* article at the very beginning that we wanted to have a "real" Farmers Market at P.F.1. This involved a series of negotiations. First, P.S.1 and MoMA told us that we should be concentrating on building the project, not making additional programming. Then, for some reason, the neighbors were worried that it would compete with another Greenmarket in Long Island City. We cleared that up. Then it took a long time to get P.S.1 and CENYC talking. Finally, we had all the permissions, a farmer willing to sell there, and we thought it was going to be great; but it wasn't.

Marcel Van Ooyen: For lack of a better phrase, you have a bunch of urban hipsters coming to listen to DJs; they weren't really interested in buying peaches while standing in line. And there was also a bit of a battle with the existing hot dog vendors. They had been there for years and didn't like us showing up; the security guards there didn't do a whole lot to help with that situation, for whatever reason. So it was an interesting concept, but, you know, live and learn.

Kate Chura: One of the things I was really happy to be able to do was to bring in Michael Pollan, the author of *Omnivore's Dilemma* and have him speak. We had him on hold to do a lecture for us, and I thought P.F.1 would be a perfect place for him to talk. We had less than a month to make this happen. I think we had two thousand people come through the door and we turned away another couple of thousand. It was just amazing how many people showed up to hear him speak, and what was fun was, it really was a combination of gardeners and artists.

Anna Kenoff: Because of the success of P.F.1, P.S.1 chose to keep the structure up a little bit longer and we created this whole harvesting event. Mike Anthony, the chef at the Gramercy Tavern who became a friend of Dan and Amale's, came and cooked a pig from Michael Grady Robertson's farm, five different ways, and he made these little veggie cocktails out of our squash and peppers and stuff. It was really exciting and drew a huge crowd.

Marcel Van Ooyen: I called it a living farm in the sky. I think it was a perfect example to show, especially to urban New Yorkers, that growing isn't beyond anyone's reach. It can be done anywhere, in any space. It basically melded a bunch of disciplines. People think farming is farming, architecture is architecture, and art is art, but this project blurred the lines between those things in ways that were really accessible—not like looking at trees behind a glass or at a picture of a farm. It was actually living and growing.

Anna Kenoff: The way that our society looks at farming and food is starting to change. Just reading what's out there in the media and different books that were popular at the moment, I think we all came to the project individually already thinking about that. And then the project as a whole really synthesized these ideas and made it real. I mean, you can read something that makes you think more about farmers' markets and your interaction with food and farming, but then to actually experience it makes it all much more feasible.

Nicholas Muraglia: Most of the people visiting were city people like me. I remember some people were almost shocked by the idea that you can eat the stuff that was growing in such an urban environment. I was with a friend and said, "You should have some of this cucumber, it's really good." And she was like, "Are you sure you can eat this?" Obviously, you wouldn't take something off the sidewalk and eat it, but in this case, in a sense, you could—erasing that space between the urban dweller and the farm and having such a direct sensual experience.

Anna Kenoff: I ate tons of vegetables that summer.

deinstall

Art Domantay: P.S.1 hired another crew for the demolition. WORKac wasn't part of the deinstall, and

I was away in Canada. I wish I could've have been there, because I really wanted to see how someone else would take it apart.

Dan Sesil: I knew it was temporary, but I was really taken with the way it was weathering. We went out a few times and sort of watched the seasoning of it, if you will. It fatigued and changed somewhat over the summer. I sort of wanted to see what nature would do to it, but they weren't going to let that happen. We always knew it was coming down, but I still would have liked to see what it looked like after the winter. It would have lasted for sure.

Mouna Andraos: Mostly what I tried to tell Sonali, who was helping with the take-down, was please go early because our stuff is tiny and fragile and we need to get it out of the way before everything else gets ripped down, so we can save some of that equipment.

Paul Mankiewicz: We took the soil back. We added a little bit of compost to it, but it was pretty much in good shape for planting. Those sixty yards went to, I think, the Davidson Avenue project up in the Central Bronx.

Amale Andraos: Sonotube is actually the largest user of recycled paper in America, so all of the cardboard was sent for recycling and probably made back into tubes. We had a guy come by and pick through it for all of the bolts and screws, which he sold for scrap metal. The plants were mostly annuals, so they got mixed in with the GaiaSoil as compost. The solar panels went back to the lenders. We still have a bunch of other equipment in the office, waiting for the next project.

Patrick Hopple: Believe it or not, when they took it apart the detailing worked. We were worried that the water would seep between the joints, but the joints held together so well that they actually squeezed the water out. The water was getting bad between the bolted areas, but where we really needed the force, it was good. We learned a lot from this. I always joke with Dan, and say, are you ready to do P.F.2?

Melani Pigat: It was successful, not because of the end result but because of how we got there—it was the people who were involved and made it happen and all of the resources that we used to research and organize. Once we finished the project, Sarah and I left pretty much the week after and never looked back; because for us, it was really about the time that we were there making it happen.

Anna Kenoff: I happened to be out of town the day that the office went to pull stuff down and I was kind of glad. Then the day that I went later to pack up stuff from the basement of P.S.1 and bring it back to the office, I didn't even go into the courtyard. I just didn't want to see it, you know? It was great while it lasted and it was nice not to have to go out there every Saturday with the watering cans, but you know, I like to remember it the way it was.

impact

Andres Lepik: For P.S.1 and for the Young Architects Program, this was really new, that a project became such a broad organization outside of the art and architecture worlds. Normally, it's a project where students are involved from the universities where the architects are teaching. But this time, it became a project with a reach far beyond all the boundaries where we normally reach museum visitors.

Sam Dufaux: P.S.1 is an incredible enclave for art in Long Island City, but somehow they never really reached the community around them. The people who live around P.S.1 are not the people who go to the museum. And I thought the farm had the effect of bringing P.S.1 to a much wider audience. The previous designs were done by architects, for architects; this time it was done for a wider community. I think that was really appealing.

Andres Lepik: I worked for fourteen years in the National Museums of Berlin, where there were a lot of discussions about what our museum audience should be. This is something I'm continuously interested in. Maybe people coming back from P.S.1 said, "Why can't we plant our vegetables on the roof or on the balcony?" I think that art and architecture can be a medium to transport other ideas to the visitors, not only that they see the Picassos and the Matisses, but that they see the museum also as a place where we can discuss other issues and think about more than just paintings hanging on the wall.

Melani Pigat: For me, the most valuable part of the whole project were all the connections and friends that we made. Not just the personal connections but, as an architecture student, learning firsthand that architects are only a small part of what makes a project really successful. That was an important lesson for me, something that can't be taught in school. Architects work with people on projects all the time—engineers, planners, developers—but this was more of a grand-scale collaboration. It was enlightening and also really fulfilling to see how excited everybody got who was involved with the project.

Art Domantay: Public Farm 1—I call it my start-up. It was the biggest project that I've done. But it was also eight years in the making. I had moved from small to medium projects in the past, and it all led up to building this P.F.1 project. I think I used every single resource and contact that I had acquired in the eight years that I had been living in Greenpoint. I loved that.

Barry Bergdoll: It was made out of cardboard and dirt and plants. And very low tech. I don't know if they planned it, but the timing couldn't have been more perfect. It went up during the period when we all believed we were still in prosperity. By the fall, Lehman Brothers and the real estate market had collapsed. So I think their project became like a sculptural symbol of a very different attitude that is just emerging. A monument, but also a debate—like, is rurbalization here? Is it coming? What is this leading to?

Michael Grady Robertson: I guess when people are enjoying their wealth in good times, they don't look around and pay as close attention to the effects it's having on nature or other people. Then, when everybody's in the same boat economically, there's a lot more care and concern for others, and a feeling of connectedness. I hope this isn't merely circumstantial, that there's something else that's driving people to look closer to home for meaning in their lives, so that even in a rebound economy, this renewed interest in agriculture and farming and community life continues.

Elodie Blanchard: The structure had a use. Compared to all those other P.S.1 projects, it was something other than building just to build something. It brought tomatoes and food—the structure was alive. You can critique any project, but that was a truly beautiful structure. That's what's so incredible, that maybe it will then give other architects, or people in the city, the idea to build a structure, make a garden. If you make it look ugly, maybe nobody would think, "Oh, yeah, we should do the same in our kids' public school or in the empty lot around the corner." If you lost the design of it, then people wouldn't get as interested, and then it wouldn't have been as good.

Andres Lepik: For us, as museum curators, we should maybe in the future look for projects that have the same impact. Not on the same level, but with the same energy and for our audience. It was interesting that the project brought so many people to P.S.1 who hadn't been there before, first-time visitors. If we can bring people to discuss art and architecture in these surroundings, when the installation is the motivation, then they see the rest as well. So, for us, it was a very positive experience to work with Dan and Amale.

Dan Wood: A year later, we received an email from MOS, the architects selected for the next installation. It was a picture of a lettuce plant growing in the P.S.1 courtyard. Apparently tomato plants have taken root in the loading dock as well!

Paul Mankiewicz: I don't think they had a beehive at Public Farm 1, so that was a serious omission; but forgivable, given that they were only there for a short while.

Michael Grady Robertson: I had a great time. That was such a unique project. I hope it will be repeated at different levels, and serve as an inspiration for other people to do work in the same vein and to learn from its successes and failures. You know, every season's sort of a new thing. I'm sure Dan and Amale know that from their first season of farming at the museum.

Lenny Librizzi: Small farms within the city limits? There are opportunities for that. There are actually two farms on Staten Island, there's the Queens County Farm

Museum, and there are the Community Gardens. There were the Victory Gardens in the First and Second World Wars, and internationally, urban agriculture is a big thing. In some African cities and in Havana, Cuba, the lion's share of any fresh produce is what people are able to grow within the city limits. There are rooftops. Vacant land. There are lots of opportunities.

Anna Kenoff: If we could have the same crew, the same team—I would definitely do it again.

THE ARCHITECT'S FARM

MEREDITH TENHOOR

Architects have long sought to improve both the food supply and social relations through designs for farms and food distribution centers. The history of this engagement is largely undocumented, but from utopian architectural farm-cities of the eighteenth and nineteenth centuries, such as Ebenezer Howard's Garden Cities, to post-World War II superfarms and markets, Archigram's hydroponic utopias, Metabolist agricultural cities, and drop-out California dome communities, the farm and the market as architectural programs have been used to stake new claims about how the urban and rural should commingle. The farm has also become a site where architects develop new concepts of sustenance and sustainability and imagine new physical connections between people and the products they consume. What follows is a very brief survey of the history of some of these undertakings.

RECONFIGURING URBAN AND RURAL

In the eighteenth century, a time of many urban food shortages, European architects sought to improve physical connections between the city and the farm. In Paris, prior to the French Revolution, with a public clamoring for cheaper bread and a city police force eager to maintain order, great architectural attention was paid to remaking the city's grain market, the Halle au Blé, where flour was sold to bakers. Though it was only a small part of the food system, the grain market was a potent public symbol of the government's commitment to filling the bellies of the populace. A new fireproof market was built in 1763. Designed by

1

2

architect Nicolas le Camus de Mézières, the market was to keep bread prices low by eliminating the overcrowding, speculation, and infernos of the old marketplace. Le Camus de Mézières began to imagine that a properly designed modern market could lower prices and actually provide better food for the politically frustrated populace. (1) Enlightenment design was put in the service of sustaining life.

German economist Johann Heinrich von Thünen was similarly fascinated with how distribution systems impacted food prices. In a series of texts published from 1818 to 1826 as *The Isolated State*, he worked out the relationship between profit and space, generating a diagram of the maximum distances that goods could travel to reach an urban marketplace.[1] Dairies and creameries tended to be located closest to cities because the goods they produced would perish quickly. Timber could be grown a bit further away from the city center, but not too far, as it was difficult to move across long distances. Farms and ranches could be located even further outside the city, since their products could survive a several-day journey.

Thünen's ideas were initially meant to describe where productive industries appeared around cities, but by the twentieth century, urban planners adopted them as an ideal model for where food industries of different types should be located.

Beginning in the late eighteenth century, French agricultural engineers were also rethinking the connections between cities and the agricultural areas that surrounded them. They realized that the human and animal waste generated in Paris would be an excellent source of fertilizer, and they proposed schemes for shipping Parisian sewage out of the city and spreading it over agricultural land. In the mid-nineteenth century, socialist Pierre Leroux developed a detailed proposal for such a system, a plan that he called the Circulus.[2] If properly channeled to the countryside through newly developed sewers, human waste could actually generate monetary value, as it would increase farm yields and return to the intestines of city dwellers in the form of fresh fruits and vegetables. City and farmland could form a closed system, doing away with the marketplace altogether.

Once political theorists saw farms and cities as interconnected systems, rather than as strictly separate entities, utopian architects and urbanists of the nineteenth century were able to creatively reconfigure the farm and the city to imagine new forms of community. Architects became interested in perfecting agricultural features as a way of engendering cooperation, self-sufficiency, and freedom from the vagaries of food prices in a speculative economy. A primary example of this kind of utopian agricultural city is French utopian socialist Charles Fourier's Phalanstère, which included ample space for cultivation. (**2**) (Fourier's work also inspired Leroux's research.) The architecture of the Phalanstère, sketched by Fourier-follower Victor Considérant, was organized radially. A central core of the complex would be dedicated to collective, quiet activities; it contained a library and meeting rooms, as well as heating systems and a communal restaurant. Extending out from the core were rooms for noisier activities, such as for educating children, as well as living quarters and workshops. Fourier situated a working farm on the periphery of the complex, where all would see it. Though no Phalanstères were actually constructed, a more compact version called Familistère was realized in 1865 in Guise, France, by Jean-Baptiste André Godin. Adapting the model to a smaller plot of land in an urban setting, the Familistère farm was replaced by a kitchen garden and a low-cost grocery store.

    Ideas for self-sustaining communities began to slip into the mainstream by the early twentieth century. Howard's classic *Garden Cities of To-Morrow* reprised the ideas of Thünen and Fourier, suggesting that an ideal city could be created by integrating all aspects of production and consumption—including agriculture—into the city.[3] (3) Since rising land prices from urban encroachment toward agricultural land made it difficult for urban farming to be profitable on the open market. Howard proposed instituting a series of financial subsidies that would make it possible to continue farming in close proximity to other urban activities. In contrast with the nineteenth-century Phalanstère and Le Camus de Mézière's grain market, Howard's ideal food-city substituted master planning for architecture. Howard was less focused on improving food distribution to curtail hunger. In his vision, agriculture was not only a means for sustaining the city's inhabitants; it generated an image of an orderly, self-sustaining, and socially cohesive community.

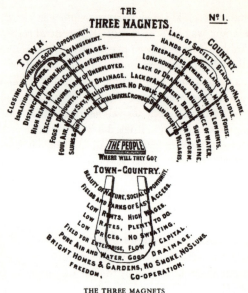

THE THREE MAGNETS

MODERNIZING THE FARM AND THE MARKET

With the spread of the railway and the perfection of water-based transportation systems in the late nineteenth century, time spent getting food to and from marketplaces shrank. Especially in the western United States, this meant that the farm could move further away from the city, and it helped excise food production to the peripheries of cities. The centers of many cities were, however, still devoted to the distribution, preparation, and storage of food. Architecture was used to promote new technologies of food preservation and distribution, which enabled food to become as much a speculative commodity as something to eat. By the early twentieth century, Dankmar Adler and Louis Sullivan were designing cold storage warehouses

171 *modernizing the farm and the market*

4

5

---

adorned with organic motifs in Chicago, and Walter Gropius had devoted great architectural attention to documenting the design of grain elevators. Soviet Constructivist architects similarly sought to promote modern agriculture with modern architecture. Projects such as architect Konstantin Melnikov's New Sukhareva Market in Moscow (1924–5) generated an image of orderly yet bountiful agricultural production that would make collective farming—something quite invisible to the average Muscovite—still apparent within the city. (**4**)

Le Corbusier's agricultural project, the Radiant Farm (1934–5) (**5**), was an outgrowth of his plan for the Radiant City. Inspired by Norbert Bézard, a rural admirer of Le Corbusier's work who challenged him to address the state of rural life in France, plans for the Radiant Farm were never realized but nonetheless represent an important attempt to imagine new forms of agricultural modernism. Taking a bird's-eye view of the situation, Le Corbusier lamented the time, land, and money lost to small-scale farming:

> From the airplane, I see infinitely subdivided pieces of land. The more modern technologies develop, the more the earth subdivides itself, and spurns the miraculous gifts of the machine. It's a total waste, it's labor frittered away.[4]

His plans for the Radiant Farm would eliminate this waste by consolidating land through the functional organization of individual agricultural plots, with zones for the production of different goods placed in close proximity to the transportation systems needed for the distribution and sale of agricultural products. Additionally, he allocated a plot of land to every citizen for kitchen gardens, which would balance the hypermodernity of the farm. He wrote:

> A farm isn't an architectural fantasy. It's something akin to a natural event, something that is like a humanized face of the earth, a form of geometric planting that is as much a part of the landscape as a tree or a hill, and as expressive of human presence as a piece of furniture or a machine.[5]

For Le Corbusier, the farm's evocation of human presence was a way to naturalize a large-scale architectural intervention in the landscape.

Efforts to modernize farm communities slowed somewhat during World War II, when industrial attention was turned elsewhere, but after the war ended, wartime technologies were repurposed to the home front and funds for economic reconstruction favored large-scale farm

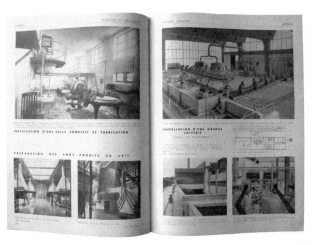

6

modernization projects grander even than those imagined by Le Corbusier. This postwar focus on industrial agriculture is somewhat odd given the popularity and success of wartime Victory Gardens, the near ubiquitous kitchen gardens that could be cultivated in urban backyards. The U.S. Department of Agriculture estimates that 40 percent of fresh vegetables in the United States were grown this way during the war. Nonetheless, planners, architects, and engineers collaborated to produce industrial-agricultural systems such as farms, markets, and distribution networks capable of providing food for growing postwar populations.

In the late 1940s, the French architectural journal *Techniques et Architecture* edited by Auguste Perret featured liberal coverage of the architecture of the industrial postwar farm. (6) New building types were surveyed, and traditional farm buildings were updated,

made subject to time and space studies to maximize their efficiency. In scale and efficiency these farms seemed to take inspiration from the Radiant Farm, but they carried few traces of its "human presence"; they were fully mechanized.

High-yield industrial farms were only part of the picture. As in the nineteenth century, architects figured out how to get products from industrial farms to markets in cities, and they advanced architectural knowledge in the process. Jean Prouvé's market in Clichy, France (1935–40), for instance, was the first building to use prefabricated curtain-wall panels. (7)

Later, Félix Candela's Coyoacán Market in Mexico City (1956) used easily constructed modular parts to quickly build structures for buying and selling. (8) Both Prouvé and Candela used the market as an opportunity to insert maximum flexibility through modern materials and prefabrication techniques. Similar strategies were applied to wholesale food markets, which received the most attention from architects. Based on models from the United States, enormous modern markets capable of feeding entire regions were constructed in Germany, the Netherlands, France, and Japan during the postwar period. They were located at the peripheries of cities, where they would easily link to highways, rail lines, and airports, and replace crowded and unsanitary inner-city markets. These markets were crucial to postwar economic development strategy: the theory went that if the middlemen and congestion costs associated with outmoded urban markets could be eliminated, food prices would be reduced for consumers, who would then have room in their budgets to purchase nonessential goods, thus stimulating economic growth.

The crème de la crème of modern wholesale markets was built just outside gastronomically inclined Paris, at Rungis from 1963 to '69. (9) Replacing the overcrowded markets at Les Halles in Paris, which once included the aforementioned Halle au Blé, Rungis became the world's largest wholesale food market.[6] Its architects—Henri Colboc and Georges Philippe—sought to lower food prices by reducing inefficiencies in the design of the market. They carefully managed the circulation of food into and out of buildings and also used computer technologies to regulate the sale prices of goods. Their efforts to improve where and how food was sold were considered to be as important as parallel attempts to modernize postwar farms, and they showed that carefully designed architecture could spur consumption of not only fruits and vegetables, but also of consumer goods like washing machines.

RETHINKING INDUSTRIAL FARMS

In spite of their appeal, standardized and industrialized farming methods, when applied too quickly, could produce disaster. In newly decolonized Tanzania, for instance, the Ujamaa agricultural villages mandated by former President Julius Nyerere from 1967–79 attempted to apply a self-generating agricultural system to the entire country. Land was collectivized, divided into a grid, and distributed presumably evenly and blindly to Tanzanian residents to grow identical crops on each plot of land. The infinite repeatability of this method of farm design was intended to reproduce social order, abundance, equality, and support for Nyerere's government, and it looked beautiful in plan. Yet in practice, plants didn't grow where they were supposed to and thousands of years of local and specific knowledge of how and where to grow food in Tanzania was lost. Though some of the Ujamaa villages were successful, technocratic planning that relied on the blind logic of the grid didn't necessarily produce high crop yields. Thousands starved under the Ujamaa system.[7]

Architectural historian Sigfried Giedion might have predicted such a disaster. In *Mechanization Takes Command*, a sort of catalog of industrial production techniques written at the behest of Walter Gropius in 1948, Giedion became fascinated with how food products became "mechanized"; that is, how the production, processing, and consumption of food increasingly depended on mechanical technologies. While for architecture, mechanization produced new forms of inhabitation and new opportunities for profit, the human mechanization (the logic of which extended directly to Nazi death camps) posed a grave danger. He wrote:

> One thing is certain. Mechanization comes to a halt before living substance. A new outlook must prevail if nature is to be mastered rather than degraded.[8]

For Giedion, a pragmatic coupling of industrial and natural agriculture was out of the question. The logic of mechanization was tight and teleological enough to preclude the coexistence of artisanal and industrial cultures; the artisanal was quickly being lost in the postwar period. As much as new technologies might be admired, their connection to living beings needed to be rethought and sometimes radically refused.

In the 1960s and '70s, Greek planner Constantinos Doxiadis followed the course of rethinking, rather than refusing. Doxiadis's approach to planning did not just consider physical infrastructure or zoning but also the organization of spaces for all aspects of life, including food production. He generated a formula for a total environment that would include space for both industrial and traditional agriculture. In his firm's journal, *Ekistics*, he argued in favor of devoting 10 percent of total land to food production areas, which he called "cultivareas." One cultivarea would be for "natural farming," filled with orchards, free-roaming animals, and open organic fields; the other would be for factory farming, with a "complete elimination of the natural landscape, where a pattern of roofs will replace a distant view of beautiful fields and orchards."[9] This technologically enhanced cultivarea would be in closer proximity to cities—a necessary evil for providing large quantities of food. In spite of his attention to the differences between these two modes of production,

10

Doxiadis didn't see that the two methods were irreconcilable; they simply served different societal functions, both of which needed to be supported. He felt that agricultural modernity could coexist with agricultural traditions.

Other architects developed more visionary and alternative food and farm schemes, attempting to resolve tensions between artisanal and industrial food production methods. Japanese Metabolist architect Kisho Kurokawa developed the Agricultural City for his first architectural project, prodigiously shown at MoMA's Visionary Architecture exhibition organized by Arthur Drexler in 1961. (**10**) Agriculture, both as a program and as a metaphor for urbanism, became a pretext for the advancement of Kurokawa's early architectural ideas. The Agricultural City was divided into cells, which could grow almost organically around an architectural framework. Raised above ground, self-generating, and non-hierarchical, the farm-city echoed the traditional development patterns of rural Japanese villages but was enhanced by modern infrastructure (in this case, streets containing utility pipes). In land-starved Japan, Kurokawa's rural plan was a surprisingly dense and urban. In a manner not dissimilar to earlier agricultural utopias, Kurokawa's settlement would enable autarchy for the farmers of individual cells. Technology could support traditional ways of life that were spatially reorganized. Like Doxiadas (but unlike Giedion) Kurokawa did not call for the radical separation of the farm from technology.

Food also became a means for thinking about the relationship between technology, consumption, and everyday life in Great Britain. In the Pop artworks of the Independent Group of the early 1950s, packaged foods appeared as ambivalent symbols of the drives toward consumption in the increasingly prosperous postwar period. Sumptuous delights—widely available thanks to improved distribution methods and the industrial farms of the immediate postwar period— became symbols of the ambivalent attitude of avant-garde culture toward the pleasures and pitfalls of a consumer society.

For the British architectural provocateurs Archigram, food's visceral symbolism became a means to tease out a new relationship between technology, political economy, and daily life. Friends with members of the Independent Group, and also featured in Drexler's Visionary Architecture show at MoMA, they used images of food and agriculture for architectural critiques. In 1963, in the third of their periodic treatises, called Archigrams (a neologism of architecture and telegram), the group expounded one relationship between food and architecture:

> Almost without realizing it, we have absorbed into our lives the first generation of expendables...foodbags, paper tissues, polythene wrappers, ballpens, EPs....
> We throw them away almost as soon as we acquire them. Also with us are the items that are bigger and last

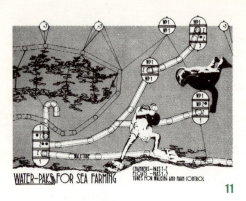

11

12

13

longer, but are nonetheless planned for obsolescence....
Our basic message? *That the home, the whole city, and
the frozen pea pack are all the same.*[10]

Like Kurokawa, Archigram was interested
in how industrial production methods, such as planned
obsolescence, could transform cities in radical ways.
Peter Cook's water paks for sea-farming (1968) showed
how food might be grown in unpredictable places. **(11)**
And in a cartoon from 1971, Cook claimed that the
farm and city could magically collapse into one another:

hologram horses and plants pop out of the roof of a
nine-story loft building. (**12**) Archigram #9 (1970), the last
of the group's dispatches, allusively and symbolically took
up the theme of cultivation. A packet of seeds was tucked
into each magazine, and the cover by Warren Chalk
featured all kinds of growing plants. For Archigram, the
plant and the farm were a symbol of a different kind of
distribution—not an idea on paper being sent by post, like
the group's then defunct magazine—but the possibility of
a living system actually being disseminated by architects
seeking to dissolve architecture into nature itself.

Technologically enhanced farms appear elsewhere
too. Hans Hollein's article "Alles ist Architektur"
(Everything is Architecture), an exposé of the increasing
range of architectural projects, features an image
of a project by architect Konrad Frey called Kuhwickel
(Cowcircle) from 1967: "A farm project incorporating
environmental control mechanisms harnessed direct to
the animals so that buildings become unnecessary."[11] (**13**)
Here the farm animal erases traditional architecture,
dissipating it into the environment at large.

RADICAL FARMING

The early 1970s produced another glut of architect-driven
farm projects, but this time they were self-commissioned,
more radical, more do-it-yourself, and were produced
by architects who dropped out rather than by famous
practitioners. Such projects were a means to rethink the
boundaries between the organic and the technological,
and a sign of participation in some of the more wild
strains of experimental architectural practice. Widespread
urban disinvestment in the 1960s created new impera-
tives for American architects to take positions about
the role of the city as a center of productivity. In the wake
of growing ecological concern and the looming oil crisis—

and after the removal of the wholesale markets
in European city centers, such as at Covent Garden in
London and Les Halles in Paris, during the late 1960s—
urban access to food and to the farmers who produce
it was made impossible within the city, and the packaged
gloss of industrialized food no longer fascinated
architects. The radical ones moved from drug-addled
cities to their own autonomous ecological compounds,
from San Francisco to Berkeley. Environmentally
friendly food emerged as a privileged site for an architect
eager to limit the alienation of modern life. Rural
England and the Pacific Northwest were ideal settings
for these new explorations, and images of geodesic dome
builders and radical communes began to proliferate
in architectural magazines.

One of the main publishers of such schemes
was *Architectural Design* (*AD*), an early champion of
Archigram and other radical practices of the early and
mid-1960s; they provided a fascinating lens for looking
at how the ideas from these practices were transformed
into food architectures. Published by the same catalog
company that printed the *Whole Earth Catalog*, *AD*
followed food out of the city, to the era's radical farms and
drop-out communities. In December 1971, *AD* published
a portrait of the Libre commune, which grew its own
food in order to become autonomous from late-twentieth-
century urban real estate speculation and technocratic
control.[12] (**14**) By 1972, food and farms were everywhere
in the magazine. The Pneumatic Issue of March 1972
featured an article called "Urban Farm." Another short
article from May 1972 featured electronic apples, which
were not intended to be eaten but rather to improve
the food supply. (**15**) Then in July of the same year,
there was an article on a Berkeley course called Outlaw
Building, which told the story of architecture students

14

15

fleeing their studios for the countryside, where they built, among other structures, orgone chicken coops, which, it was believed, would restore the birds' vital sexual energy for the production of more eggs.[13] Adjacent to this little article was another about electromagnetic gardening, a technique of applying radiation to vegetable gardens to increase their yields.

In the first article, students who have been toiling "under fluorescent lights" were invited to go to the country so that they could "grow under the sky." In the second, radiation was considered more valuable and less dangerous, at least for vegetables; it could

16

enhance rather than destroy them. While seemingly endorsing the benefits of natural living in rural areas, it's clear that these articles also had an agenda for architecture that is not dissimilar to Frey's. They transformed the food environment into an expanded field for technologically dependent architectural production. Not even the counterculture's loud cries for natural and organic living could keep food from its technologized fate in *AD*. In such a context, one could run away from fluorescent lights and eat electromagnetic vegetables without a conflict.

As the oil crisis hit in 1973, a rural escape to a self-sufficient community seemed like less of a radical gesture and actually became somewhat pragmatic. At this time, *AD* began to feature more articles on growing food within cities, using high-technology methods appropriated from engineering periodicals. The Circulus reappeared in Reyner Banham's sketches for an ecologically controlled environment that converts household waste into energy. (**16**) The January 1974 issue featured articles on vertical growth systems and crops that could be grown for energy rather than food. There were highly technical discussions of the potential for extracting energy from plants, which could be "applied at a community level" for both sustenance and electricity. In his book *The Ecological Context* (1970), *AD* contributor John McHale similarly posits protein as a pure source of life, a quantifiable commodity whose consumption varies according to the level of privilege and exploitation practices of different continents. Here, food is oddly divorced from any pleasure one might take from eating. The body, experiencing the ecological crisis, eats chemical formulas, not cake. By the end of the period, architecture was no longer even necessary, only farming. You could even "grow your house."

These projects were small interventions, experiments, and fantasies. They took the threats to late-capitalist culture posed by the energy crisis seriously, and tried to picture a reordered world, whether it be the desirable and pleasurable one of the Libre commune or novelist Ernest Callenbach's *Ecotopia* or the strangely dystopian and likely tasteless one of the portable fish farm designed for high protein yields, featured in *AD* of 1971. But these projects also put forth an architectural agenda. They ambitiously expanded the concept of what architecture did and what materials it could be made from: seeds, plants, soil, fertilizer, electromagnetism, geodesic domes, energy-transforming equations, inflatable structures, and clothing, not to mention experimental forms of communal life. While it was clear that architecture could pose no real solution to the complex problems associated with ecological collapse, these small examples seemed like ingredients for change, both architectural and ecological. But when cheap oil was restored in the 1980s, the profession and its magazines largely returned to work on organizing spaces for speculation and consumption, such as shopping malls, markets, and museums. Architectural periodicals stopped documenting wild experiments in ecological living.

OBLIQUE FARMING

Today there's an incredible awareness of the environmental, health, and social dangers of industrial meat and grain production, rampant and untested genetic modification, and the global proliferation of agricultural monopolies. As a result, the utopian back-to-the-land DIY communes of the early 1970s are slowly reappearing in the mainstream architectural press. There is a new appetite for the pure products of small organic farms and a clamoring for an end to government policies that make it difficult for these farms to survive. Gastronomes and ecologues alike are rethinking not only how food is produced, but also how it might be better distributed from rural areas to cities. Community Supported Agriculture (CSA) programs sell subscriptions of produce to city dwellers, providing stable income to family farms; for example, New York's New Amsterdam Market (an initiative to build a permanent indoor market for locally and sustainably produced food) focuses on increasing profits for farmers, not lowering food prices for consumers. Thriving and convivial communities are often assembled around such initiatives, yet forming them is highly inefficient when viewed through the industrial lens. These homegrown strategies are "horizontal"; that is, they take time and energy and require an abundance of land that only seems to be available in rural areas. In contrast, increasing global urbanization over the past twenty years has helped to generate a multitude of "vertical" strategies for growing food inside of the city's limits. This interest in urban food is due in part to the second oil crisis and a growing awareness of the social and environmental costs of shipping food long distances.

187 *oblique farming*

17

18

But it is also due to the fact that urban land is now highly valuable and desirable, and property owners, architects, and planners have conspired to rethink how food might be grown in cities, given these new economic and ecological pressures. From techno-futurist urban farms such as MVRDV's Pig City, which squeezed a pig farm into a skyscraper (**17**), to Dickson Despommier's Vertical Farm, which proposes growing hydroponic vegetables in the curtain-wall of a skyscraper, architects have designed vertical farms not only to make cities more livable, pleasant, and sustainable, but also to make cities into better sites for investment and real estate profit. Even engineering giant ARUP has been researching how to grow food—organic, no less—within cities and has produced a demonstration of their methods in plans for a new "eco-city" in Dongtan, China. (18)

But aren't there more transformative ways that architects can use farms today? What would happen if architects did what they have always done: design a better means of navigating the politics and techniques of contemporary food production? It's clear from this history that the relationship between two strains of farming that we're now familiar with—industrial versus organic, horizontal versus vertical—actually become less diametrically opposed when they enter the hands of creative architects. P.F.1 may be such an example. Like Kurokawa's Agricultural City, it's neither horizontal nor vertical, but rather a farm that's "oblique" (to borrow Paul Virilio and Claude Parent's term). As Amale Andraos has said, "P.F.1 only doubles up the ground enough to create a social space underneath the farm." Unlike massive, high-tech vertical farms, P.F.1 seems to point out that urban space can also support small-scale farms that generate communities around the production of food. And unlike the hypothetical or remote farms appearing on the pages of *AD* in the '70s (which are impossible to really digest unless you like cellulose and toxic ink!), thousands of people could viscerally experience the pleasures of P.F.1 by eating the food it produced. At P.F.1, WORKac assembled both cardboard *and* communities. The time-consuming and strange work of actually making (and not just drawing or photomontaging) small utopias—assembling armies of volunteers, managing the complexity of such an undertaking, and getting more people to taste and dance under a canopy of food that is grown with love and effort—is properly architectural work.

1. Johann Heinrich von Thünen, *Isolated State: An English edition of Der isolierte Staat*, ed. Peter Geoffrey Hall, trans. Carla M. Wartenberg (Oxford: Pergamon Press, 1966).

2. See Dominique Laporte, *History of Shit*, trans. Nadia Benabid and Rodolphe el-Khoury (Cambridge: MIT Press, 2000).

3. Ebenezer Howard, *Garden Cities of To-Morrow (Being the Second Edition of "To-Morrow: a Peaceful Path to Real Reform")* (London: S. Sonnenschein & Co., 1902).

4. Le Corbusier, "La 'Ferme Radieuse', le 'Village Radieux'," in *L'Homme Réel 4*, (April 1934), 54–59. (Excerpt translated by Meredith TenHoor)

5. Ibid., 59.

6. See Meredith TenHoor, "Architecture and Biopolitics at Les Halles," *French Politics, Culture and Society* 25, no. 2 (Summer 2007): 73–92.

7. See James Scott, *Seeing like a State: How Certain Schemes to Improve the Human Condition Have Failed* (New Haven: Yale University Press, 1998). Also see Zaki Ergas, "Why Did the Ujamaa Village Policy Fail?—Towards a Global Analysis," *Journal of Modern African Studies* 18, no. 3 (September 1980): 387–410.

8. Sigfried Giedion, *Mechanization Takes Command: A Contribution to Anonymous History* (New York: Norton, 1969), 256.

9. C. A. Doxiadis, *Ecology and Ekistics*, ed. Gerald Dix. (London : Elek, 1977), 22.

10. *Archigram* 3 (Autumn 1963).

11. Hans Hollein, "Alles ist Architektur," *Bau: Schrift für Architektur und Städtebau* 20, no. 1/2 (1968): 1–28; Hollein's article was republished as "Alles is [*sic*] Architecture," *Architectural Design* 2 (1970): 62; the article in *AD* was the first English translation.

12. "Libre," *Architectural Design* 41 (December 1971): 727–36.

13. "Outlaw Building," and "EMG," *Architectural Design* 42 (July 1972): 399.

AFTERWORD
PRAGMATOPIA?

WINY MAAS
IN CONVERSATION WITH DAN WOOD & AMALE ANDRAOS

WINY MAAS: For me, there are two components to your P.F.1 project: the content, which I think is tremendous; and the installation itself, the tubes that are glued together in the wave shape, over the public space. From a design point of view, it is very nice and also ironic—with all those pots up in the air. I wonder, though, whether that "wave" is a bit overaesthetic and becomes more formal rather than productive or functional. So, that's funny. Does it mean that the wave itself got more attention? Maybe. Although the fact that you did it like that has had a better impact on the audience, especially the New York audience.

DAN WOOD: That is actually something we always worry about—but from the opposite perspective. We are always concerned that, by focusing on concepts and ideas, we get pigeonholed as "nondesigners."

AMALE ANDRAOS: Right. When we did P.F.1, we brought this new agenda to a competition that had previously been purely about form and design. So it's funny to hear you think that maybe it was too designed. I think everybody focused on the content as if there was no design at all!

WM: You know, we all love food these days. Urban culture is so connected to eating and all of the aspects surrounding it. There's a kind of sexiness to food that deserves attention. For me, it's almost awful that our food production is sometimes positioned far away, in circumstances that are, let's say, nonsexy, and that are highly distant, not only in physical but also in psychological terms. Food and farming should be closer entwined with urban elements.

To see the project as a component of a larger study—figuring out which part of the food supply urban farming can provide—is good. But it's a complex study. On the one hand, it's good to reduce the distance, in terms of food production, for ecological reasons. On the other hand, if you extend that thought, it might lead to the autarky of cities; that could kill, for example, the complete production of the Valle Central in Chile. So the entire breadth of North American trade would need to be repositioned, and I think we don't yet have a full overview of the consequences.

AA: For us, here in the United States, the continent is so big and industrial farming has become such a disaster that we feel that you can still be quite radical and yet still have a net positive effect.

WM: In Europe, there is a big discussion about buying wine from Chile or Australia. Even if they are fantastic wines and organically grown, there is still discussion because they are produced so far away. But the fact is, the climate in certain parts of the planet is supergood for farming and might be better for a lot of products than the zones around the world's bigger cities. That's worth a discussion.

DW: Definitely. It links back to the systems that support cities. Does it make sense for every building to have individual solar panels, or should we harness the best zones for solar and then bring the energy back to the city? It's the same with food. It doesn't make sense to grow food everywhere; certain areas are much more productive. It's a balance between the local and the logical.

WM: I try to explain it in our book, *KM3*, excursions on capacities in one of the "ideal worlds." If we put everything on the best spot, climate-wise—solar energy in the Sahara and the west of the United States, and food production in the moderate climates—then yes, that "world" is actually quite good, because there is codependency and a balance. But we have to solve the transport issues for both solar energy and food products.

AA: The question for us, as architects and urban thinkers, is how can we really engage these issues in a constructive way? What is a possible role of the visionary today?

WM: What is the definition of visionary? There is, today, a tendency to think: visionary equals large-scale equals experiments from the past (that were not completely successful). People are therefore highly suspicious of the word visionary. But if we define visionary in new terms, say, in terms of leadership or progressive suggestions for the future, then suddenly there is more appeal. In this highly individualized world, leadership is a relief sometimes and responsibility has become an obligation for everyone because of the condition of the climate. So there is a paradox between that suspicion of the visionary and the need for leadership. Crisis helps. Everyone today wants to know a little bit more about the future; therefore, the visionary has potential again.

Architecture and urbanism are extremely useful tools to discuss what should be done. Because we can visualize, we can show the effect of thoughts on space. That gives the profession a strength that has been neglected in the "formal days," as I call them, of the '90s and the beginning of this decade, when the architecture debate was toward exuberance in form. In that respect, yes, I think visionary architecture is again possible.

AA: How is this visionary moment different from the kind of utopias and dystopias of the past? We are looking for ways that the visionary, today, can be more progressive, specific…pragmatic, maybe. What are the tools to avoid this pendulum from utopia to dystopia?

WM: The question is how to develop that dichotomy and turn it into a more productive situation. Indeed, the pragmatic, as you said, helps. The fact that many of the components that can lead to visionary architecture or urbanism are actually based in other domains, like infrastructure, social equality, or ecology, helps as well. That allows the architectural utopia to avoid becoming completely hermetic and distant for the wider audience.

DW: Engaging in cross-disciplinary thinking and working, with real expertise, rather than just architects projecting.

WM: Exactly. The architect does not want to become, in the example of your P.F.1, an agrarian economist, but the architect can visualize what agricultural economists are working on, and, by visualizing, criticize—confronting these fields with practical, buildable issues and inspiring perhaps new agricultural economic models.

AA: As architects, we can make things visible, and by making them visible, we add value to something that could have functioned in the same way, but invisibly.

WM: But that approach on its own does run the risk of becoming purely educational. We need to take architecture a step further, to do something with it, make another twist. I think the word evolutionary comes in. If we add an evolutionary layer to that educational one, then maybe we will start to find a way to deal with that critique. What I mean is that, when you go back to a building, you see how it has changed over time: gets updated, richer, grown-up. That's what I would like to add to your suggestion regarding visualization.

You know William McDonough's lecture, where he shows this image of a bird's nest and eggs on the green roof and he says, "That's architecture"? I love it. It's such a good one-liner. But then I wonder are there always birds, are the birds still there? Or does it even become an ecological disaster, too many birds of the same kind, for instance, which are killing all the insects in that area? I'm intrigued by the evolutionary aspect of that roof. And so I'd ask them to show the next slide; let's give it more fruitful thought, like, what can help?

DW: One of the criticisms of modernism that happened in the '70s was its overfunctionalism. Everything became a formula, and it was often inhuman. And now the ecological, which is supposed to be the "warm and fuzzy" antithesis to modernism, is becoming too formulaic and functional. It's an interesting reversal.

WM: It should not just be about solving things, but raising things to a higher level. Functionalism kills curiosity sometimes. It kills invention. For example, when you look at 90 percent of the recent LEED buildings in the United States, they are extremely ugly. So it's interesting to ask, why did these buildings become so ugly? Is it because they were not inventive, they didn't choose, they didn't make it sexy, they didn't make it social—or whatever? How can we give the ecological agenda more content?

There is another question around this issue. Just like the term visionary needs a translation for these days, the word ecology does as well. Some people say that ecology is everything and includes human beings in a complete system, while others only concentrate on one element, like energy or carbon dioxide. I'm of a tendency to give professions their own place. Sociology, for example, is a profession and is not ecology, therefore.

AA: We agree. In theory, ecology may be everything, but that's not a very useful definition to work with. So we also tend to bring it back to specific systems that can be measured and used in combination with architecture and urbanism. Ecology becomes a kind of productive added value.

WM: The good thing about ecology these days is that it gives an agenda, and almost no one is against it. Of course, that agenda reaches back to the 1970s, with the Club of Rome. And before that, perhaps even Le Corbusier was already visionary in that respect, Ebenezer Howard as well.

In time, though, this subject has been deepened and more fully worked out. There's more technology and more awareness. At the beginning of the twentieth century, it was more connected with enlightenment and a small group of theoreticians dealing with this new architecture. Now, ecological thinking has a wider impact. It's nice to see the snowball effect.

APPENDIX A
WARM VEGETABLE SALAD

MICHAEL ANTHONY

This salad is based on a principle that drives our food philosophy at Gramercy Tavern: the layering of light, seasonal flavors. We frequently repeat the same ingredient, handling it in different ways, be it raw, pickled, glazed, or puréed, creating an echoing effect for a particular season. This dish can be enjoyed bite by bite, savoring each ingredient, or devoured by the spoonful.

(Yield: 6–8 servings)

SALSIFY
2 LARGE Salsify, peeled and cut into 1" lengths
2 CUPS Vegetable stock
1 TBSP Butter
1 TBSP Yellow clover honey
1 TBSP Garlic oil
3 TBSP Ginger juice
TO TASTE Salt and pepper

Combine all ingredients in a sauce pan and cook until a syrupy glaze forms. Set aside.

ASPARAGUS TIPS
1 BUNCH Asparagus, tips only
2 CUPS Vegetable stock
1 TBSP Butter
1 TBSP Yellow clover honey
1 TBSP Garlic oil
3 TBSP Ginger juice
TO TASTE Salt and pepper

Combine all ingredients in a sauce pan and cook until a syrupy glaze forms. Set aside.

BLACK RADISH

1 LARGE Black radish
¼ CUP Beet juice
1 CUP Raspberry vinegar
2 CUPS Vegetable stock
1 TBSP Sumac
3 SPRIGS Thyme
3 CLOVES Garlic cloves, peeled
1½ TBSP Butter, unsalted
1 TBSP Yellow clover honey
TO TASTE Salt and pepper

Cut black radish into ⅛" slices. Using a ½"-diameter ring mold, cut rounds from each slice. Combine radish rounds with the remaining ingredients and cook until reduced to a syrupy glaze. Set aside.

CARROTS

2 BUNCHES (APPROX. 12) Thumbelina carrots, peeled and cut in half lengthwise
2 CUPS Vegetable stock
1 TBSP Butter
1 TBSP Yellow clover honey
1 TBSP Garlic oil
3 TBSP Ginger juice
TO TASTE Salt and pepper

Combine all ingredients in a sauce pan and cook until a syrupy glaze forms. Set aside.

BABY TURNIPS

2 BUNCHES (APPROX. 20) Baby turnips, well scrubbed and trimmed
2 CUPS Vegetable stock
1 TBSP Butter
1 TBSP Yellow clover honey
1 TBSP Garlic oil
3 TBSP Ginger juice
TO TASTE Salt and pepper

Combine all ingredients in a sauce pan and cook until a syrupy glaze forms. Set aside.

PICKLED CHARD STEMS

- **4 CUPS** Swiss Chard stems, cut into batonnet
- **2 CUPS** Sugar
- **2 CUPS** Water
- **6 CUPS** Rice wine vinegar
- **2 CUPS** Salt
- **½ TSP** Mustard seed
- **½ TSP** Black peppercorns
- **½ TSP** Fennel seed
- **½ TSP** Coriander seed
- **1 EACH** Red beets, peeled and chopped

Mix all of the above ingredients, except the chard stems, and bring to boil. Remove from heat and strain. Let cool. Add chard stems. Refrigerate overnight.

SUNCHOKES

- **2 LARGE** Sunchokes, peeled and cubed
- **2 CUPS** Vegetable stock
- **1 TBSP** Butter
- **1 TBSP** Yellow clover honey
- **1 TBSP** Garlic oil
- **3 TBSP** Ginger juice
- **TO TASTE** Salt and pepper

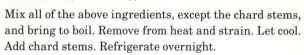

Combine all ingredients in a sauce pan and cook until a syrupy glaze forms. Set aside.

BEETS

- **1 BUNCH** Red baby beets
- **1 BUNCH** Yellow baby beets
- **1 BUNCH** Candy cane baby beets
- **3 TBSP** Butter, unsalted
- **TO COAT** Olive oil
- **1 CUP** Water
- **3 TBSP** Salt
- **1½ TBSP** Pepper

Preheat oven to 350°F. Coat each color beet seperately with olive oil, salt, and pepper and place each in its own baking pan with remaining ingredients. Cover with foil and bake 45 minutes. Remove; when cool enough to handle, peel beats. Trim and cut into quarters. Set aside.

CANDIED LEMON ZEST

Lemon rind, cut into strips
Simple syrup
Water

Place lemon rinds in cold water and bring to a boil. Strain. Place lemon rinds in simple syrup and bring to a simmer 10 minutes. Remove from heat.

LEMON-FENNEL PURÉE

1 OZ Garlic cloves, cut julienne
1 OZ Shallots, peeled, and cut julienne
½ LB Fennel, cut julienne
A PINCH Saffron
6 PIECES Candied lemon zest
1 LEMON Lemon juice
TO TASTE Simple syrup
1 TBSP Olive oil
TO TASTE Salt and pepper

Sweat garlic, shallots, and fennel in olive oil until soft and translucent. Add saffron, lemon zest, lemon juice. Cook 10 minutes more. Season with simple syrup, salt, and pepper. Place mixture in blender and purée.

LEMON VINAIGRETTE

2 CUPS Lemon juice
2 CUPS Lemon oil
3 TBSP Wild flower honey
3 TBSP Onion puree
1 CUP Lemon-fennel purée
3 TBSP White wine vinegar
3 TBSP Olive oil (Divina)

Combine lemon juice, honey, and white wine vinegar. Whisk and slowly drizzle in lemon oil to form an emulsion. Whisk in olive oil. Add onion purée, lemon fennel purée, and honey; mix until well combined. Season with salt and pepper.

BEURRE BLANC

- **1 CUP** Champagne vinegar
- **1 CUP** White wine vinegar
- **½ CUP** White wine
- **1 EACH** Bay leaf
- **2 SPRIGS** Thyme, picked
- **1 TSP** Coriander seed
- **1 TSP** Fennel seed
- **1 EACH** Shallot, peeled, and cut julienne
- **1 EACH** Garlic clove, cut julienne
- **¼ CUP** Heavy cream
- **¾ LBS** Sweet butter

Place liquids with herbs, spices, garlic, and shallots in a pot. Cook until reduced. Add heavy cream and reduce again. Whisk in butter by the tablespoon until melted.

ADDITIONAL VEGETABLES

- **1 EACH** Radicchio trevisano, tips only
- **1 EACH** Sunchoke, peeled and thinly sliced into ribbons
- **3 LEAVES** Swiss chard, cut into 1" squares
- **4 EACH** Easter egg radishes, cleaned, trimmed, and sliced thin
- **3 CUPS** Farro, cooked al dente

YOGURT WALNUT DRESSING

- **1 CUP** Plain yogurt
- **1 CUP** Mixed herbs, chopped: parsley, tarragon, dill
- **½ TBSP** Parmesan cheese, shaved
- **1 ½ TBSP** Walnut oil
- **1 ½ TBSP** Olive oil (Riviera)
- **½ TBSP** Cilantro syrup
- **½ TBSP** Whole toasted walnuts
- **TO TASTE** Salt
- **1 EACH** Lemon juice

TO ASSEMBLE

Toss farro (p. 200) with beurre blanc (p. 200) and lemon vinaigrette (p. 199) just to coat. Season with salt and pepper. Arrange cooked vegetables over and around farro; then place raw vegetables over and around farro. Drizzle with yogurt walnut dressing (p. 200).

Michael Anthony joined Gramercy Tavern as the executive chef in 2006. Under his leadership, the restaurant has earned a number of accolades, including a three-star New York Times *review in 2007 and the James Beard Foundation Award for Outstanding Restaurant in 2008. In both 2008 and 2009, Anthony was nominated for the James Beard Foundation Award for Best Chef: New York City.*

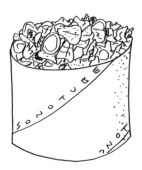

APPENDIX B
URBAN CHICKENRY

HOW TO KEEP A CHICKEN COOP

PREPARATION

Be prepared. Check your local municipal ordinances and follow the law.

Stay informed. Do research on best practices and methods for keeping your new chickens happy, healthy, and safe.

KNOW YOUR CHICKENS

They like to sleep and roost in elevated places. Make a perch. There should be about 10 inches of perch per bird.

Chickens need 4 to 10 square feet of space per bird in the coop and 4 to 8 square feet in the run.

Hens like to lay their eggs in safe, dark places. Give them a designated nest area.

Chickens often get bored. Make sure to change their environment frequently.

They like to take dust baths to get rid of mites and to stay cool. Give them access to a patch of soil.

FOUR BASIC TYPES OF COOPS & FOUR ELEMENTARY COMPONENTS

COOP WITH PEN
This coop is ideal for a backyard with ample space.

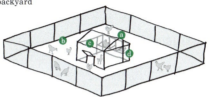

COOP WITH RUN
A chicken run can replace the pen if space is slightly restricted.

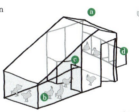

CHICKEN TRACTOR
Distribute chicken manure fertilizer by moving the coop around on wheels.

CHICKEN ARK
This coop is compact and, like the chicken tractor, can be moved around a yard quite easily.

COMPONENT KEY
a. shelter
b. exterior space
c. perch
d. nest

P.F.1 CREDITS

ARCHITECTURE & CONSTRUCTION
WORK Architecture Company
PRINCIPALS:
Amale Andraos and Dan Wood
DESIGN TEAM LEADERS:
Anna Kenoff and Haviland Argo
DESIGN TEAM:
Jenny Andersen, Haviland Argo, Fred Awty, Sarah Carlisle, Diego Chavarro, Sam Dufaux, Katherine Eberly, Gaustas Eigirdas, Morten Federsen, Julia Galeota, Alina Gorokhova, Anna Kenoff, Tamicka Marcy, Heidi Cathrine Østergaard, Melani Pigat, Bryony Roberts, Elliet Spring, Samuel Stewart-Halevy, David Peterson, and Magda Szwajcowsk
SITE LEADERS AND VOLUNTEER COORDINATORS:
Sarah Carlisle and Melani Pigat
CONSTRUCTION VOLUNTEERS:
Zehra Ahmed, Kyle Barker, Aria Bilal, Kiki Blazevski, Hendy Bloch, Jennie Boija, Clark M. Buccola, Kimberly Cases, Jane Chen, Jimi Chen, Karin Chen, Jeff Chinn, Luke Cho, Amanda Chong, Christina Ciardullo, Scott Del Rossi, Douglas Philip Dunlap, Lisa Ekle, Mirian Elrassi, Rafael Enriquez, Lian Eoyang, Matthew Eshleman, Gabriele Falconi, Melissa Goldman, Zhidong Gong, Deborah Grossberg, Frank Guitard, Anders Gurda, Jose Gutierrez, William Hood, Mark Horne, David Huber, Grace Hwang, Diane Jacobs, Emily Johnson, Gregory Katz, Mi Hyun Kim, Nathan King, Anna Komocka, Joseph Littrell, Laura Liu, Candice Luck, Alison Macbeth, Ruth Mandl, Eliza Montgomery, Sunmi Moon, Nickolas Muraglia, Miki Onodera, Sonia Parada, Daniel Payne, Sandra Perez, David Quintana, Lakhena Raingsan, Ashley Reed, Soojung Rhee, Julia Rice, Lizzie Rothwell, Emily Scarfe, Lawrence Siu, Gaines Solomon, Parima Sukosi, Gianne Sultana, Grace Tang, Denver Thomas, Alexandra Tiffin, Ryan Welch, and Jeffrey Yip

CONSTRUCTION
Art Domantay
COORDINATION:
Art Domantay
TEAM:
Phyllis Baldino, Kenneth Grady Barker, Jake Borndal, Ross Chambers, Tim Daly, James W. Dugdale, Dave Giansante, Adam Golovoy, David Haiderzad, Byron Johnston, Wade Kavanaugh, Alberto Lopez, Hugo Mendoza, Chris Morgan, Stephen Nyugen, Yayoi Sakurai, Miguel Santana, Matt Suter, Jules Swyers, Eric Thorne, and Marie Ucci

PLANTING CONSULTATION & GREENHOUSE SPACE
Queens County Farm Museum (QCFM)
Amy Fischetti and Michael Grady Robertson

P.F.1 credits

STRUCTURAL ENGINEERING
LERA
Daniel A. Sesil, Matthew Melrose, and Patrick Hopple

FABRIC CONSULTATION & DESIGN
elasticCo.
Elodie Blanchard

ELECTRONICS, SOUND, & VIDEO
Electronic Crafts
Mouna Andraos

SOUNDSCAPE DESIGN & FARM SOUNDS
Michael J. Horan

GRAPHIC DESIGN
Project Projects
Prem Krishnamurthy, Adam Michaels, Neil Donnelly, Kevin Wade Shaw, Molly Sherman, and Chris Wu

GREENHOUSE / GREENTEAM
The Horticultural Society of New York (HSNY)
Kate Chura, Fiona Luhrmann, and Hilda Krus

RAINWATER COLLECTION, FARMERS' MARKET, & GENERAL SUPPORT
Council on the Environment of New York City (CENYC)
Marcel van Ooyen, Lenny Librizzi, Lars Chellberg, and Julia Leung

SOLAR POWER CONSULTATION & INSTALLATION
altPOWER
David Gibbs, Andy Allbee, and John Foster

PLANT AND SOIL SUPPORT & IRRIGATION
Town and Gardens

GAIASOIL ULTRA-LIGHTWEIGHT GROWING MEDIUM
The Gaia Institute
Paul S. Mankiewicz and John Halenar

GARDENING CONTAINERS
Smart Pots

FERTILIZER
Terracycle

CARDBOARD TUBES
Sonoco
Wim van de Camp and Dee Welsh

CNC MILLING
Associated Fabrication
William Moawat, Amy Stringer, Jeffrey Terras, and Kenneth Tracy

IRRIGATION SYSTEM
Rain Bird
Dave Shane

IRRIGATION CONSULTANT
Atlantic Irrigation Specialties
Matt Hart

SOLAR PANELS
Atlantis Energy Systems

INVERTER AND SOLAR CHARGE CONTROLLERS
OutBack Power Systems

SEEDS
Burpee
George Ball, Jr.

BATTERIES
Power Battery

T-SHIRTS
American Apparel

REPROGRAPHICS
Digital Plus
George Uenishi

OTHER VOLUNTEERS
Green Apple Core and the Center for Architecture Foundation's High School Students Program

DONOR & VOLUNTEER
Freshfields Bruckhaus Deringer US LLP

DONORS
The New York Water Taxi
The Seed Fund
MensVogue.com

P.F.1 SPECIAL THANKS
Louis & Anne Abrons Foundation Inc., All-City Metal, Stan Allen, American Apparel, George Ball, Benjamin Ball and Gaston Nogues, Shigeru Ban/Dean Maltz, BACO Enterprises, The Brightman Hill Charitable Foundation, Compostwrks, Robert Fogelson and Victoria Voytek, Amy Falder, Freshfields Bruckhaus Deringer US LLP, G & C Crane Service, Marty Gottlieb, Fritz Haeg, Evan and Regina Haymes, Sharon Coplan Hurowitz and Richard Hurowitz, Maria Kane, Richard and Gail Maidman, Marjam, *Men's Vogue*, New York Water Taxi, Dolly Mirchandani, Poly Pro Industrial Coatings, Deedie and Rusty Rose, Rust-Oleum, Marc Simmons and Elaine Yau, Eric Stein and Rebecca Olshin, The Seed Fund of the Studio for Urban Projects, Pat Tarrant, and Leon Yaloz/Steve Goldman/Java Street LLC

INVENTORY BOOKS SPECIAL THANKS
Shannon Harvey, Prem Krishnamurthy, Kevin Lippert, Laurie Manfra, Irvin Michaels, Margaret Michaels, Noah Sheldon, Molly Sherman, Jennifer Thompson, Deb Wood, Chris Wu

IMAGE CREDITS

Unless otherwise noted, all images
© WORK Architecture Company.

Cover: Raymond Adams
p.29: Jacob Silberberg
pp.56–57, 75, 90, and 93:
Raymond Adams
pp.129, 132–135, 137, and 139:
Elizabeth Felicella
pp.140–1: Mikkel Bøgh
pp.142–4: Noah Sheldon
p.148: Raymond Adams (left);
Elizabeth Felicella (right)
p.154: Raymond Adams
p.164: Hilary Sample
p.167: Bibliothèque nationale
de France (1, 2)
p.170: Ebenezer Howard (from
Garden Cities Of To-Morrow,
published by MIT Press)
p.171: Princeton University Press
p.173: *Techniques et Architecture*
p.174: Centre Pompidou, Paris,
Bibliothèque Kandinsky (7);
Reinhold Publishing Corp. (8);
Rungis International (9)
p.178: Kisho Kurokawa
Architect & Associates
p.180–3: *Architectural Design*
p.184: Getty Research Institute
p.187: MVRDV (top);
ARUP International (bottom)

Published by
Princeton Architectural Press
37 East Seventh Street
New York, New York 10003

For a free catalog of books,
call 1.800.722.6657.
Visit our website at
www.papress.com.

© 2010 Project Projects LLC
All rights reserved
Printed and bound in China
13 12 11 10 4 3 2 1 First edition

No part of this book may be used
or reproduced in any manner
without written permission
from the publisher, except in the
context of reviews.

Every reasonable attempt has
been made to identify owners of
copyright. Errors or omissions will
be corrected in subsequent editions.

Special thanks to: Nettie Aljian,
Bree Anne Apperley, Sara
Bader, Nicola Bednarek,
Janet Behning, Becca Casbon,
Carina Cha, Tom Cho, Penny
(Yuen Pik) Chu, Carolyn
Deuschle, Russell Fernandez,
Pete Fitzpatrick, Wendy Fuller,
Jan Haux, Erin Kim, Linda
Lee, John Myers, Katharine
Myers, Steve Royal, Dan Simon,
Andrew Stepanian, Jennifer
Thompson, Paul Wagner,
Joseph Weston, and Deb Wood
of Princeton Architectural Press
—Kevin C. Lippert, publisher

Series editor: Adam Michaels
PAP editor: Laurie Manfra
Design: Project Projects

Project editor: Heather Peterson
Design and coordination
assistance: Kevin Wade Shaw

The editors gratefully acknowledge
the generous support of the Graham
Foundation for Advanced Studies
in the Fine Arts.

Library of Congress
Cataloging-in-Publication Data

Above the pavement—the farm! :
architecture & agriculture at
PF1 / Amale Andraos and Dan
Wood, editors.—1st ed.
    p. cm.— (Inventory books)
    Includes bibliographical
references.
    ISBN 978-1-56898-935-8
    (pbk. : alk. paper)
1. Work Architecture Company.
2. Urban agriculture—New York
(State)—New York. I. Andraos,
Amale, 1973– II. Wood, Dan, 1967–
III. P.S. 1 Contemporary Art Center.
IV. Title: Architecture & agriculture
at PF1. V. Title: Architecture and
agriculture at PF1.
    NA737.W67A84 2010
    728'.92—dc22
                      2009046213